SpringerBriefs in Environmental Science

SpringerBriefs in Environmental Science present concise summaries of cutting-edge research and practical applications across a wide spectrum of environmental fields, with fast turnaround time to publication. Featuring compact volumes of 50 to 125 pages, the series covers a range of content from professional to academic. Monographs of new material are considered for the SpringerBriefs in Environmental Science series.

Typical topics might include: a timely report of state-of-the-art analytical techniques, a bridge between new research results, as published in journal articles and a contextual literature review, a snapshot of a hot or emerging topic, an in-depth case study or technical example, a presentation of core concepts that students must understand in order to make independent contributions, best practices or protocols to be followed, a series of short case studies/debates highlighting a specific angle.

SpringerBriefs in Environmental Science allow authors to present their ideas and readers to absorb them with minimal time investment. Both solicited and unsolicited manuscripts are considered for publication.

More information about this series at http://www.springer.com/series/8868

Mathew Kurian · Reza Ardakanian
Linda Gonçalves Veiga · Kristin Meyer

Resources, Services and Risks

How Can Data Observatories Bridge
The Science-Policy Divide
in Environmental Governance?

 Springer

Mathew Kurian
UNU-FLORES
Dresden
Germany

Reza Ardakanian
Office of the Director
UNU-FLORES
Dresden
Germany

Linda Gonçalves Veiga
School of Economics and Management
University of Minho
Braga
Portugal

Kristin Meyer
UNU-FLORES
Dresden
Germany

ISSN 2191-5547   ISSN 2191-5555 (electronic)
SpringerBriefs in Environmental Science
ISBN 978-3-319-28704-1   ISBN 978-3-319-28706-5 (eBook)
DOI 10.1007/978-3-319-28706-5

Library of Congress Control Number: 2015960237

Printed on acid-free paper

This Springer imprint is published by SpringerNature
The registered company is Springer International Publishing AG Switzerland

The original version of this book was revised:
The copyright holder name has been changed.
The erratum to this chapter is available
at 10.1007/978-3-319-28706-5_5

Preface

Data, its generation, collection, sharing and analysis, and its power to influence decision making and support-coordinated action in support of policy goals are what drives the functionality of the Nexus Observatory. The potential applicability of the Nexus Observatory as a tool for agenda setting and monitoring progress in sustainable management of environmental resources provided the rationale for establishing the Africa Consortium on Drought Risk Monitoring. Through a focus on risk, it was possible to convince relevant ministries, nongovernmental organizations (NGOs), and donor agencies of the need to address issues of infrastructure operation and maintenance to support the delivery of critical public services such as water supply and irrigation. By collaborating with European universities, it also becomes possible to combine in situ data collection by regional partners with the power of remote sensing and earth observations to enable data analytics employing multiple mediums including mobile and GIS. In addition to supporting the development of robust feedback loops between science and policy, we hope such an endeavor will also identify opportunities for strategic engagement with the policy process based on analysis of cases of "success" and "failure" in international development.

UNU-FLORES, in its role as a think tank of the United Nations system, recently established the Nexus Observatory initiative to inform discussions on feedback loops and knowledge translation. The Nexus Observatory is an online platform that hosts *inter alia* databases, an online learning portal and dedicated data sets that rely on a consortium comprising UN agencies, member states, and regional universities and training institutes. The Nexus Observatory consortia demonstrate that feedback loops are important in highlighting the relationship between individual behavior, resource allocation by public agencies, and environmental outcomes. The scientific robustness of the initiative can be gauged by the extent to which regional consortia can calibrate their response to the impact of global changes such as urbanization, climate, and demography while accommodating for trends such as decentralization and the emergence of information and communication technologies (ICTs) that have had a discernible impact on governance structures and processes.

This volume provides the theoretical basis for pursuing the idea of a Web-based observatory that addresses the science-policy divide in environmental governance. We posit that the absence of disaggregate, reliable, and frequent information at appropriate scales makes it difficult to predict the environmental outcomes of infrastructure construction. Moreover, the absence of regional capacity to collect, analyze, and transmit information to decision makers curtails the ability of governments to respond to disaster risks effectively. As a consequence, the possibility of establishing a robust system for monitoring international development goals (e.g., sustainable development goals) is curtailed.

We have organized the volume into four chapters that demonstrate the need for a perspective that treats environmental resources, the services they support, and the risks that disasters pose to effective delivery of services in a holistic manner. This would enable us to reflect critically upon the strength of the poverty–environment nexus while guiding us with the design of programs and projects focused on addressing the challenges of environmental sustainability. In constructing our argument, we draw upon five cases from Philippines, India, Laos, and Honduras to elaborate upon five divides that characterize environmental governance today: (1) infrastructure versus services, (2) centralized versus decentralized government, (3) public versus private management models, (4) short-term versus reliance on long-term planning perspectives, and (5) efficiency versus equity. The cases we draw upon cover water, soil, and waste resources; services; and associated disaster risks.

Mathew Kurian

Acknowledgements

This Springer Brief is based on an analysis of case studies covering the nexus of water, soil, and waste resources. The arguments presented in the brief benefitted from comments received on a paper presented at the first Joint UNU-FLORES and TUD seminar in Dresden, Germany, a Nexus Observatory proposal writing workshop hosted by Institute for Global Environmental Strategies in Tokyo, Japan; a panel discussion at Dresden Nexus Conference 2015 on data, monitoring, and governance involving Nexus Observatory consortium partners from Asia and Africa; and a regional consultation on the nexus approach to management of environmental resources that was hosted by the Water Development Management Institute (WDMI) of the Ministry of Water of the United Republic of Tanzania in Dar es Salaam. Leslie O'Brien (our language and copy editor) supported the preparation of the manuscript to enable its final submission in compliance with Springer guidelines.

Contents

Chapter 1
Introduction to the Volume: A Strong Case for Data and Observatories in the Context of the Post-2015 Monitoring Agenda

Half a century of development experience has pointed to several shortcomings in our approach to agenda setting. A stark reminder of one of those shortcomings is reflected in theory that purported to view development as progress along pre-determined *stages of growth*. Jagdish Bhagwati of Columbia University was quick to point out some of the contradictions of international aid based on a simplistic reading of how developing societies produced, saved, and invested their output (Bhagwati 2010). A simplistic reading of development led to a disproportionate emphasis on creation of infrastructure (*outputs that included dams and water treatment plants*). The post-2015 monitoring agenda will hopefully reflect the great distance we will have come from when we emphasized outputs to the present time when there has been a discernable shift in emphasis toward development outcomes (*water quality and agricultural productivity*) and impact in terms of human well-being and environmental sustainability (United Nations 2015).

In our most recent book, *Governing the Nexus*, the United Nations University (UNU-FLORES) has documented case studies and global experience to argue that the science-policy divide is at the heart of the disconnect between development and achievement of outcomes and impact both in terms of human well-being and environmental sustainability (Kurian and Ardakanian 2015a). We articulated the science-policy divide in the form of the following question: Why does statistical significance not always correspond with political action. Our research leads us to conclude that there are two important mismatches between the research enterprise and public policy formulation. The first is the absence of robust feedback loops. The second is a lack of imagination about the modalities concerning knowledge translation (Kurian and Ardakanian 2015b).

UNU-FLORES, in its role as a think tank of the United Nations system, recently established the Nexus Observatory[1] initiative to inform discussions on feedback loops and knowledge translation. The Nexus Observatory is an online platform that hosts *inter alia* databases, an online learning portal and dedicated data sets that rely on a consortium comprising UN agencies, member states, and regional universities and training institutes. The Nexus Observatory consortia demonstrate that feedback loops are important in highlighting the relationship between individual behavior,

---

[1]https://nexusobservatory.flores.unu.edu/.

M. Kurian et al., *Resources, Services and Risks*, SpringerBriefs in Environmental Science, DOI 10.1007/978-3-319-28706-5_1

resource allocation by public agencies, and environmental outcomes. The scientific robustness of the initiative can be gauged by the extent to which regional consortia can calibrate their response to the impact of global changes such as urbanization, climate, and demography while accommodating for trends such as decentralization and the emergence of Information and Communication Technologies (ICTs) that have had a discernible impact on governance structures and processes (Kurian and Meyer 2015).

The Nexus Observatory initiative also underlines the importance of knowledge translation. Far from being a linear process involving uptake of scientific output, decision-making may entail having to "muddle through" based on important political trade-offs that may neither promote equity nor efficiency goals. While scientists endeavor to achieve precision with their results, an effective bridge to the policy domain could strive to do more in terms of making trade-offs explicit. This change in approach has several implications. First, it means we acknowledge the significance of decentralization—political, fiscal, and administrative—and its potential to affect decisions and development outcomes at scale. Second, it means that once trade-offs are made explicit, individuals and public agencies will be motivated to design incentives that foster synergies that address common challenges such as water scarcity or pollution. Third, for solutions to emerge, data that is reliable, frequent, and sufficiently well disaggregated is important to ensure that decision makers can predict the scale and intensity of the policy challenge and bring to bear a proportionate amount of human and financial resources to realize the achievement of clearly verifiable development outcomes and impact.

Data, its generation, collection, sharing and analysis, and its power to influence decision-making and support coordinated action in support of policy goals is what drives the functionality of the Nexus Observatory. The potential applicability of the Nexus Observatory as a tool for agenda setting and monitoring progress in sustainable management of environmental resources, provided the rationale for establishing the Africa Consortium on Drought Risk Monitoring. Through a focus on risk, it was possible to convince relevant ministries, Non-Governmental Organizations (NGOs), and donor agencies of the need to address issues of infrastructure operation and maintenance to support the delivery of critical public services such as water supply and irrigation. By collaborating with European universities, it also becomes possible to combine in situ data collection by regional partners with the power of remote sensing and earth observations to enable data analytics employing multiple mediums including mobile and GIS. In addition to supporting the development of robust feedback loops between science and policy, we hope such an endeavor will also identify opportunities for strategic engagement with the policy process based on analysis of cases of "success" and "failure" in international development.

This volume provides the theoretical basis for pursuing the idea of a web-based observatory that addresses the science-policy divide in environmental governance. We posit that the absence of disaggregate, reliable, and frequent information at appropriate scales makes it difficult to predict the environmental outcomes of infrastructure construction. Moreover the absence of regional capacity to collect, analyze, and transmit information to decision makers curtails the ability of governments to respond to disaster risks effectively. As a consequence the possibility of establishing a robust system for monitoring international development goals (e.g. Sustainable Development Goals) is curtailed.

We have organized the volume into three chapters that demonstrate the need for a perspective that treats environmental resources, the services they support, and the risks that disasters pose to effective delivery of services in a holistic manner. This would enable us to reflect critically upon the strength of the poverty-environment nexus while guiding us with the design of programs and projects focused on addressing the challenges of environmental sustainability. In constructing our argument, we draw upon five cases from Philippines, India, Laos, and Honduras to elaborate upon five divides that characterize environmental governance today: (1) infrastructure versus services, (2) centralized versus decentralized government, (3) public versus private management models, (4) short-term versus reliance on long-term planning perspectives, and (5) efficiency versus equity (Kurian and Ardakanian 2015b). The cases we draw upon cover water, soil and waste resources, services and associated disaster risks.

References

Bhagwati, J. (2010). Banned aid: Why international assistance does not alleviate poverty. *Foreign Affairs, 89*(1), 120–125.

Kurian, M., & Ardakanian, R. (Eds.). (2015a). *Governing the nexus: Water, soil and waste resources considering global change*. Dordrecht: UNU-Springer.

Kurian, M., & Ardakanian, R. (2015b). The nexus approach to governance of environmental resources considering global change. In M. Kurian & R. Ardakanian (Eds.), *Governing the nexus: Water, soil and waste resources considering global change* (pp. 3–13). Dordrecht: UNU-Springer.

Kurian, M., & Meyer, K. (2015). *The UNU-FLORES nexus observatory and the post-2015 monitoring agenda*. Retrieved from https://sustainabledevelopment.un.org/content/documents/6614131-Kurian-The%20UNU-FLORES%20Nexus%20Observatory%20and%20the%20Post-%202015%20Monitoring%20Agenda.pdf

United Nations. (2015). *Global sustainable development report: 2015 edition, advanced unedited version*. New York: United Nations Department of Economic and Social Affairs, Division for Sustainable Development.

Chapter 2
Institutions and the Nexus Approach

Trade-Offs, Synergies and Methods for Governance of Environmental Resources

Abstract This chapter draws upon three published case studies covering different aspects of water, soil, and waste management to discuss issues that are crucial to an improved understanding of the nexus approach to management of environmental resources. Based on previous research, we identified three questions that can guide the discussion on the nexus approach (Kurian and Ardakanian in Governing the nexus: water, soil and waste resources considering global change, UNU-Springer, Dordrecht, 2015). (1) Question of intersectionality: What are the critical mass of factors at the intersection of material fluxes, public financing and heterogeneity, and changes in institutional and biophysical environment that define environmental outcomes? (2) Question of interactionality: How can feedback loops be structured to capture both vertical and horizontal interactions between legal and policy reform, structural changes in economy and society, and variability in the biophysical environment? (3) Question of hybridity: What role can transdisciplinary methods play in supporting integrative analysis of biophysical and institutional processes that have a bearing on the use and management of environmental resources?

Keywords Data · Monitoring · Environmental resources · Risks · Governance · Public services · Case studies · Nexus observatory · Index · Visualization · Benchmarking · Scenario analysis · Trade-offs

1 Introduction

"The Nexus Approach to environmental resources' management examines the inter-relatedness and interdependencies of environmental resources and their transitions and fluxes across spatial scales and between compartments. Instead of just looking at individual components, the functioning, productivity and management of a complex system is taken into consideration" (UNU-FLORES 2015). The emphasis on moving away from a focus on "individual components" is evident from the Integrated Water Resources Management (IWRM) approach. The IWRM approach emphasizes the following issues: (1) Inter-sectoral competition for surface water

M. Kurian et al., *Resources, Services and Risks*,
SpringerBriefs in Environmental Science, DOI 10.1007/978-3-319-28706-5_2

resources; (2) Integration of water management at farm, system, and basin scales; (3) Conjunctive use of surface and groundwater resources; and (4) Prioritizing water for human consumption and environmental protection (Turral 1998; Schreier et al. 2014). However, some have pointed out in previous analyses that IWRM approaches neglect the political dimension through a reification of "natural boundaries" and emphasis on "neutral planning and participation" (Wester and Warner 2002: 65).

The novelty of the nexus approach to management of environmental approaches is that it considers the above critique of IWRM seriously by engaging with scientifically problematic concepts of *trade-offs* and *synergies* (Kurian and Ardakanian 2015). The concept of a trade-off by signifying a compromise involves negotiation. Similarly, the concept of a synergy by signifying inter-connection necessitates collective action. What portfolio of mixed methods can capture the relationships between naturally available environmental resources (*flows and fluxes*), the demands that are made on them as a result of resource extraction strategies (*budget allocation*) of public agencies and individual behavior of resource users (*example: crop choice*), and environmental outcomes (*example: soil erosion or water quality*) (Poteete et al. 2010)? This question underscores the need for integrated analysis of *resources, services, and risks,* which constitute important components of poverty and well-being (Dasgupta 2001). Without a robust analytical framework, the science-policy divide will continue to support a disconnect between development and achievement of outcomes in terms of well-being and/or environmental sustainability (Kurian and McCarney 2010).

Advancing the nexus approach to management of environmental resources necessitates the development of a methodological framework that integrates analysis of biophysical processes with political structures and behavior that shapes environmental outcomes (Veiga et al. 2015; Reddy et al. 2015). Further, the nexus approach demands the delineation of *boundary* and *scale* conditions to be able to distinguish between the effects of human intervention and ecological change on management outcomes (Scott et al. 2015). The example of groundwater is illustrative: the objective of watershed management is to facilitate recharge of aquifers for which a number of human interventions are possible, such as construction of water harvesting structures or delineation of safe zones. After a few years, recharge of aquifers may become evident, but the important question is how much of this change has resulted from management when an improvement in rainfall or precipitation patterns during the same period could have also played a crucial role (Kurian 2010).

This chapter draws upon three published case studies covering different aspects of water, soil, and waste management to discuss issues that are crucial to an improved understanding of the nexus approach to management of environmental resources. Based on previous research, we identified three questions that can guide the discussion on the nexus approach (Kurian and Ardakanian 2015). (1) *Question of intersectionality:* What are the critical mass of factors at the intersection of material fluxes, public financing and heterogeneity, and changes in institutional and biophysical environment that define environmental outcomes? (2) *Question of interactionality:* How can feedback loops be structured to capture both vertical and horizontal interactions between legal and policy reform, structural changes in economy and society, and variability in the biophysical environment? (3) *Question of hybridity:* What role can transdisciplinary methods play in supporting integrative analysis of biophysical and institutional processes that have a bearing on the use and management of environmental resources?

The first case study employed by this chapter focuses on soil and water management in the Mekong river basin of Laos. The second case study examines the issue of wastewater reuse for peri-urban agriculture in India. The final case study examines the issue of "collective action" through longitudinal analysis of self-organization for management of watershed resources in the Indian Himalayas. The three case studies highlight issues of complexity, heterogeneity, and uncertainty, as well as the trade-offs and synergies that make robust management regimes possible. The case studies also elaborate upon the methodological innovations that are required to enhance the credibility of research results for decision makers.

2 Understanding the Nexus: Approach, Concepts and Methods

2.1 Approach: Inter-dependence, the Commons and Collective Action

The nexus approach highlights the importance of inter-dependence involving environmental resources. Concepts of system fluxes and flows between compartments and across land uses capture the aspect of inter-dependence. However, inter-dependence could also be extended to include interactions across resource management boundaries potentially covering rural and urban, central and provincial governments and departments. We have argued previously that the literature on new-institutional economics uses the concept of *externalities* to elaborate upon the complex issues relating to human and agency behavior, and inter-dependence (Kurian 2010). As an example, the rapid expansion of piped and non-piped water supply resulted in exponential rates of wastewater generation. In many instances, wastewater in a developing country context is discharged into water bodies—rivers or groundwater aquifers. Rivers and groundwater are characterized as Common Pool Resources (CPRs) because they are *non-excludable* and *non-rival* in nature. This is very much like the argument that purported that "before Chlorofluorocarbons were invented, the stratospheric ozone layer was a public good; and since it was provided by nature, there was no under-provision. Now it is a CPR, subject to human depletion" (Keohane and Ostrom 1995: 15). Public choice theory has argued forcefully that it may indeed be possible to reverse the "degradation trap" we face by designing robust institutions[1] that can be effectively aligned horizontally (*involving*

---

[1]Institutions are understood here as rules covering potentially constitutional, collective choice and operational rules (Ostrom 1990).

departments such as forestry or agriculture) and vertically (*involving ministries such as finance or community irrigation management cooperatives*) (North 1990; Ostrom 1990). Collective action that emerges when rules are properly aligned and monitored over time and space, can result in effective management of multiple environmental resources and uses (Kurian and Dietz 2013). Robust collective action strategies in turn may advance principles of subsidiarity, accountability, and autonomy (Veiga et al. 2015).

2.2   Key Concepts

The nexus approach emphasizes the fact that effective management of environmental resources can in many instances necessitate the mitigation of trade-offs. For example, an afforestation program to restore soil fertility at basin scale may increase uncertainty in terms of downstream water flows. Further, increasing spatial and temporal scale may break the causality between poverty and environmental outcomes because of the amplification of effect of exogenous factors such as seasonal differences in farming practices, fluctuations in factor and product markets, and divergence in strategies of extension agencies (Kurian and Ardakanian 2015: 4–5). Some have referred to this as a Poverty-Environment (P-E) nexus (Lal 2015). Depending on the strength of the causal relationship between poverty and environment, public programs may have different impacts. For instance, a sector-wide approach assumes that the P-E nexus is strong while a budget support approach assumes the P-E nexus is weak (Dasgupta et al. 2005).

New Institutional Economics (NIE) has also been instrumental in shifting the focus away from a world of only two possibilities—centralized/hierarchically organized public sector departments and privatization of publicly owned assets with the aim of improving service delivery (Kurian 2010: 10). In this context, scholars have drawn upon public choice literature to introduce the concept of co-production defined "as a process through which inputs used to produce a good or service are contributed by individuals who are not in the same organization" (Ostrom 1996: 1074). Ostrom's analysis of condominial sewers in Brazil illustrates issues that pertain to the role of community participation for management of environmental resources:

1. Centralizing infrastructure provision at the national level keeps local governments from access to decision-making responsibilities;
2. Excessively high engineering standards by distant bureaucracies were inadequate for bringing better services to poorer regions and neighborhoods; and

3. Citizens were themselves helpless to do anything about squalid conditions although they possessed resources like skills and time.

Co-production as a concept can help improve our understanding of the design of interventions that aim at improving delivery of public services such as irrigation or water supply. Here NIE scholars have argued that effective co-production involving public and private sectors and individual citizens can result from a multi-level and nested framework of rules (Ostrom 2009). Ostrom provides an overview of this framework by delineating four levels of rules, the proper alignment of which necessitates synergies:

1. Resource systems (*example, covering specific territory containing a water system*);
2. Resource units (*example, amount and flow of water*);
3. Governance systems (*example, organizations responsible, specific rules related to use of environmental resources and how these rules are made*); and
4. Users (*individuals, households, groups who use an environmental resource*).

A major insight offered by Ostrom's framework relates to the nesting of rules: "All rules are nested in another set of rules that define how the first set of rules can be changed" (Ostrom 1990: 51). Whenever one addresses questions of *institutional change,* it is essential to recognize the following: Changes in rules used to order action at one level (example, community water user association) occur within a currently "fixed" set of rules at a higher level (for example, constitution). Further, changes in higher-level rules are usually more difficult and costly to accomplish when compared to collective choice and operational rules.

The literature on New Institutional Economics has made an important distinction between institutional environment—policy and legal environment (*including property rights*)—and institutional arrangements. Institutional arrangements refer to access to markets, information, technology, financial resources, and skilled staff with clearly defined roles and responsibilities. The dispersion of institutional arrangements will most certainly differ across space (village, town or watershed). Here, *process variables* like connectivity to critical infrastructure such as roads, power or internet or motivated agency staff may mediate access to institutional arrangements. Interestingly, realization of higher order service outcomes like delivery of affordable and reliable water/wastewater services are often mediated by market forces. In this context, availability of disaggregated data can hamper planning aimed at improving access to basic services relating to water and soil resources (Kurian 2010). The big institutional question that can be posed is: How are resource allocation practices (transfers, taxes, tariffs, and functions and functionaries) advancing optimization of environmental resource use? An examination of this question has implications for design of research covering methods for data generation, collection, sharing, analysis, decision-making, and coordinated action (Kurian and Turral 2010). This is why transdisciplinary methods offer a number of benefits, an issue we examine in greater detail in the ensuing discussion.

2.3   Studying the Nexus: The Role of Transdisciplinary Methods

"Transdisciplinarity is used for research that addresses the knowledge demands for societal problem solving regarding complex societal concerns" (Hirsch Hadorn et al. 2006: 122). Because the nexus approach to the management of environmental resources typically includes such complex, inter-dependent processes and approaches to problem solving, transdisciplinary methods can assist with advancing the nexus approach and bridging the gap between science and policy.

Transdisciplinarity is characterized by crossing disciplines, scales, boundaries, and geographies as well as sectors, space and time. "As the prefix *trans-* indicates, *transdisciplinarity* concerns that which is at once *between* disciplines, *across* different disciplines, and *beyond* all disciplines. Its goal is understanding the present world and the unity of knowledge" (Bertea 2005: 6). Compared with other forms of research methods, such as interdisciplinarity or multidisciplinarity, transdisciplinarity introduces a new openness and coordination between hierarchical levels of science and sources of knowledge, including institutional perspectives (Max-Neef 2005). In this way, as argued by Bertea, transdisciplinarity moves beyond specialization in isolation, also hyper-specialization (disciplinarity), multiple disciplinary perspectives of a given theme (multidisciplinarity), and hierarchical research coordination (interdisciplinarity). It rejects an artificial separation of knowledge, instead permitting a holistic assessment (Bertea 2005) that is reflective of reality[2] and integrates rational thought (the true or false paradigm) with relational thought (the right or wrong paradigm) (Max-Neef 2005; Lang et al. 2012). Additionally, transdisciplinary research helps with giving structure to complexities of societal challenges by considering four levels of knowledge/evidence in both their horizontal and vertical interrelations and interdependencies. These levels consist of the following: (1) The empirical level (what exists), (2) The pragmatic level (what is possible), (3) The normative level (what is desired/needed), and (4) The purposive/value level (what should be done and how) (Hirsch Hadorn et al. 2006; Max-Neef 2005).

---

[2]Reality includes a reflection of the diversity, complexity and dynamics of related processes, as was highlighted by Hirsch Hadorn et al. (2006).

In doing so, transdisciplinary methods attempt to overcome hierarchies of knowledge, fragmentation, and silo mentalities. This has also been emphasized by Gibbons et al. (1994: 163), who proclaim: "Though scientists remain the driving force in proposing areas for research, the research priorities will be generated within hybrid fora composed of many different actors".

Any research process that engages actors from various backgrounds (scientists, policy makers, practitioners, civil society, etc.) and subsequently competing interests, approaches, and methods, must address a number of challenges throughout the research process. Brandt et al. are among those, who have identified key challenges that may jeopardize the success and impact of transdisciplinary research. Points of friction may arise from a lack of coherent research frameworks, non-integration of diverse methods, differences concerning research and knowledge production processes, as well as the nature of involvement of non-scientists, such as policy makers or practitioners. Additionally, the generation of impact is hampered by the lack of platforms for publication of results, both in the field of science and practice (Lang et al. 2012).

The abovementioned challenges to employing transdisciplinary methods bear certain implications for research that advances a nexus approach to the management of environmental resources. As Lang et al. concluded, a structural change enabling transdisciplinary research is a crucial step toward increasing collaboration, integrating the best available knowledge, and co-producing knowledge that offers solution options to real-world problems (Lang et al. 2012).

By taking a problem solving, collaborative approach attitudes may be transformed, competences, capacities and ownership developed, institutions reformed, and technologies developed (Hirsch Hadorn et al. 2006). This also involves depicting knowledge sites (e.g. local or scientific knowledge), different policy cultures as well as their interactions and power relations with regard to the production of knowledge and consequently the participation of relevant stakeholders (Gibbons et al. 1994; Pohl 2008; Lang et al. 2012; Jahn 2008). "We need to focus on the quality of the engagement process, and move away from reliance on using a science disciplinary filter to judge the quality of the information or knowledge we use" (Apgar et al. 2009: 269). By extension, this assertion raises questions as to where knowledge is situated and, subsequently, stimulates debates concerning social accountability frameworks and feedback loops, including social learning, environmental management, and participation.

By employing a hybrid form of research that is based on mixed methods, complex societal problems can be addressed more coherently. Transdisciplinary methods, thus, constitute a key step toward providing a more complete evidence-base when addressing nexus challenges. Most importantly, it forces researchers and other stakeholders involved in the process to look beyond their own interests and instead to rethink their approaches to knowledge generation, problem solving, social learning (especially between those involved), and means of communication.

3   The Nexus Approach to Management of Soil, Water and Waste Resources: What Lessons from Case Studies?

3.1   Case Study 1: Soil and Water Conservation in Laos[3]

3.1.1   Introduction

In recent years, significant reductions in levels of poverty in Vietnam, Lao PDR, and Cambodia have been accompanied by a shift toward market and trade liberalization (Kurian and Lestrelin 2004). Fallow periods under traditional slash and burn system of cultivation have been shortened as a result of changes in land tenure policies (Evrard and Goudineau 2004). One particular policy of relevance to the discussion relates to Article 17 of the Land Law of Laos. Article 17 stipulates conditions for allocation of land to farmers, but does not make allowance for individual family size. The decline in fallow periods however has increased weed infestation, heightened labor requirements, and intensified tillage practices in upper catchments of the Mekong river basin. The transformation of the slash and burn system has exacerbated soil erosion and reduced soil fertility and crop yields with adverse consequences for livelihoods of poor farmers in upper catchments (Roder 1997; MRC 2003). This case study, by reporting on the findings of the Management of Soil Erosion Consortium (MSEC) research project that aimed to identify technical options for soil conservation, focused on two research questions: (1) What are the potential socio-economic impacts of adoption of alternative soil conservation practices and management systems in the Mekong river basin? (2) What are the policies, legal, and organizational constraints to adoption of alternative technical options for soil conservation measures?

3.1.2   Ex-post Evaluations: The Role of Transdisciplinary Methods

It is important to understand the relationship between human behavior (e.g. *land use, payments for services, crop or technology choices*) and their impact on the condition of environmental resources (*water, soil, waste*). Analysis of MSEC experimental field trials revealed that farmer adoption of technical options such as *improved fallow*[4] were yet to take place. We argued instead that preliminary results of studies on crop yields and soil and nutrient loss, when analyzed in tandem with results of studies on cost-benefit and crop production, can highlight potential for

---

[3]See Kurian (2010) for more information on this case study.

[4]Improved Fallow (IF) option includes planting of pigeon pea and crotolaria seeds in plots of annual crops like maize, Job's Tears and upland rice to enrich the poor bush fallow with additional biomass, early ground cover and extra litter to improve soil and suppress weeds in a short period (De Rouw et al. 2003: 17).

technology adoption by farmers (Kurian 2010: 198). We argue that the absence of data on actual adoption rates should not prevent attempts to understand the *potential* for technology adoption.

Studies on technology adoption based on time series data may improve our understanding of biophysical processes (erosion, nutrient load) but does not necessarily enrich our understanding of institutional factors that mediate technology adoption by farmers. Scott rightly points out, "schemes for introduction of new crops such as cotton, tobacco, groundnuts and rice as well as plans for mechanization, irrigation and fertilizer regimes were preceded by lengthy technical studies and field trials. Why, then, have such a large number of these schemes failed to deliver anything like the results foreseen for them" (Scott 1998: 288)? Scott suggests that the failure lies partly in the obsession with data generated by a narrow, experimental, and exclusively quantitative approach that drives out other forms of local knowledge and judgment posed by cultivators (Kurian 2010).

We envisage three particular benefits of prospective studies in supporting transdisciplinary approaches to bridging the science-policy divide. First, social learning based on consultations with farmers' groups and field staff of parastatal agencies could enhance processes of information exchange, capacity building, and project cycle management (Biggs and Smith 2003). Second, even as technology development gathers pace, a number of institutional issues related to labor availability for farm operations and access to markets for agricultural inputs and commodities are flagged. Finally, a methodology that incorporates perspectives on distribution of costs and benefits of technology adoption among farmers of differing class and gender categories could highlight pro-poor impacts of technology adoption across different spatial and temporal categories (Kurian and Dietz 2005).

In our analysis of soil conservation in the Mekong river basin, we found that it was imperative that we measured gross margins and costs that farmers experience from crop production under a given set of technological practices. In our case study location, we found that farmers do not incur costs on fertilizers and transportation. Therefore, gross margins we ascertained were most likely influenced by per ha yields and labor costs. Others have pointed out that daily wages for agricultural labor are normally paid at the rate of 4–6 kg of rice plus free meals (Roder 1997: 4). Therefore, in calculating labor costs, we assumed that every employable member of a household would offer his or her services on farms of at least three households. Depending on family size of the household hiring out labor, the same household can expect to receive three times the same number of labor to help with farm operations. This figure when added to the number of members of employable age in the household would give us the total number of laborers available to perform farm operations free of charge. We assumed that every additional laborer required would incur a wage of US$1.50 per day.

Transdisciplinary perspectives also influenced our approach to calculating yields for Job's Tears under the improved fallow system. In this case, we allowed for a 20 % increase in per ha yields over the slash and burn system for plots in higher reaches of the catchment, a 30 % increase for plots in the middle reaches of the catchment, and a 40 % increase for plots in lower reaches of the catchment. These categories fit well with the range of figures of yield increase for Job's Tears as reported by agronomic studies in Houay Pano catchment (see De Rouw et al. 2003). Potential constraints to generating gross margins were size and quality of farm plots, quality of output (*for example rice or Job's Tears*), and differences in price received by farmers for sale of agricultural crops.

Our analysis of food security benefited from nutrition studies in Lao PDR. We found that an average person would consume 535 g of rice daily yielding approximately 1900 kcal, which gives a 90 % adequacy of daily intake (Aree et al. 2003: 10). Sen (1999: 81) makes a strong methodological case for assigning evaluative weights to different components of quality of life (forests, soils or household incomes) and then to place the chosen weights for open public discussion. Others emphasized the need to examine issues of intergenerational equity in tandem with generating and analyzing data on biophysical change to understand potential for farmer adoption of alternative technologies and management practices (see Dasgupta et al. 2005). Based on group discussions, we found that balanced nutritional requirements come normally from consumption of vegetables, crabs, rats, and forest products. The population at the study site consumed meat, eggs, and poultry products less frequently. By combining information on household size with 1999 market price for a kilo of rice, we calculated household food requirements. We then calculated the extent to which gross margins derived from sale of Job's Tears would meet household food sufficiency requirements of poor households.

We found that on average, each household comprises three members of employable age. The availability of members of employable age within households is crucial in determining potential for cultivation of commercial crops like bananas and pineapples. In the case of pineapples, two years after initial planting the crop is available for harvest annually in August. However, since the weeding operations for pineapple coincides with similar operations for rice, Job's Tears, and maize, there is intense demand for farm labor. Most households prioritize achieving household food security based on production of rice and maize or sale of Job's Tears to purchase food items. On average, we found that each household could sustain its food production from domestic production for up to nine months. Therefore, using nine months as an average, we found that about 30 % of households did not meet the average figure and we classified them as food insecure.

3.1.3   Issue of Trade-Offs in Management of Soil Resources

Erosion rates are higher especially for short rotation systems under the traditional slash and burn agriculture system practiced in large parts of the Mekong river basin. Interestingly, agronomic studies in our study area revealed that although erosion rates were highest under the slash and burn system, crop yields were highest under this system, averaging between 1900 kg per ha for rice and 1400 kg per ha for Job's Tears. By contrast, crop yields of Job's Tears under the improved fallow system (*known to lower erosion rates*) averaged only 400 kg per ha (Kurian 2010: 202). Another alternative to the slash and burn system, known as mixed cropping of Job's

Tears and Pigeon Pea (*with no tillage*), caused a yield reduction of between 26 and 53 % (De Rouw et al. 2003). This analysis highlights the importance of trade-offs between efficiency and conservation that operates at the plot or farm scale.

Transdisciplinary research approaches improve our understanding of trade-offs in environmental management. We found in our discussions with farmers the importance of topographic features like slope and soil erosion in influencing crop yields on individual plots. For example, on erosion plots that are located on steep plots, Job's Tears yields were more likely to be in the range of 400–500 kg per ha under the slash and burn system. Job's Tears yields under the slash and burn system were likely to be still lower and in range between 200 and 300 kg per ha respectively on erosion prone plots that are located on land with moderate to lower slopes. Therefore, an improved fallow system has the potential to increase per ha agricultural yields on erosion prone plots with diverse slope characteristics (Kurian 2010: 204).

Improved fallow system has the potential to improve per ha yields and gross margins from production of Job's Tears. However, another trade-off needed examination: Would improved fallow be socially and politically acceptable as a technological alternative? In other words: 1. How would adoption of improved fallow technology affect the distribution of gross margins from production of Job's Tears among different categories of farmers and 2. How would adoption of improved fallow technology affect food security among different categories of farming households? It was revealed that farmer adoption of improved fallow technology had the potential to reduce levels of income poverty. Gross margins from production of Job's Tears increases by US$73 and US$33.20 for middle and low-income households respectively. As a result, the number of farmers in the low-income category declines since gross margins from production of Job's Tears using improved fallow has the potential to move them into a middle-income category. Transdisciplinary approaches to data collection and analysis revealed that for households most likely to improve their food security, intra-household decision-making related to food and expenditure patterns would ensure that benefits would be distributed more evenly irrespective of gender or age status.

3.1.4 The Institutional Environment and Scope for Synergies in Environmental Management

Transparent policy processes as reflected in information exchange between MSEC scientists and field staff of parastatals were assumed to have an important influence on farmer adoption of alternative practices at different scales: experimental plot, farm, catchment, watershed or river basin. The predictability of the legal and policy framework as reflected in, for instance, stability of market prices for some crops or flexibility of land-tenure reform process could also influence farmer adoption of alternative practices. The sensitivity of cropping patterns to changes in market prices was evident, for example in the 1990s, when the decline in price of coffee on the international market resulted in a sharp decline in the area under coffee cultivation in Luang Phabang from 313 ha in 1990 to 90 ha in 2000 (State Planning

Committee 2000). It is also important to recognize that fluctuation in export volumes of agricultural crops has been accompanied by a steady devaluation of the Laos currency (*Kip*) against the US Dollar between 1999 and 2004.

Our case study revealed the importance of institutional environment (*policy and legal framework*) and institutional arrangements (*markets, labor, land*) in influencing environmental management (Kurian 2010: 205). With regards to institutional arrangements, three factors were highlighted by our analysis.

1. Rigid land tenure reform process apparent from Article 17 of the Land Law makes no allowance for changes in plot size due to population growth. Rigidity of the land-tenure process constrains farmers from benefit of the potential yield increases offered by alternative technological practices.
2. Poor access to information on sources of good quality seeds affects quality of crop production reflected in different prices that farmers received for their output. Further, inaccurate information on market conditions was responsible for government extension workers persuading farmers to cultivate Job's Tears as a cash crop. However, prices fluctuated due to regional constraints as private companies reneged on their purchase agreements with farmers. This left farmers with insufficient cash to make food purchases.
3. Technical capacity of the government extension agencies and departments affected the quality and reliability of advice farmers received on seeds and crop varieties. Limited access to nationwide data sets on issues such as trends in land cover change limited the capacity of local government departments to identify resource management priorities and plan future investments.[5] Adequate staffing levels were another issue that constrained the ability of government departments in undertaking effective management of environmental resources.

3.2   Case Study 2: Wastewater Reuse in Peri-Urban Agriculture in India[6]

3.2.1   Introduction

One of the consequences of urbanization is land modification. Land modification combined with gradient dynamics, can alter the intensity and direction of water flows in urban areas. Wastewater usually enters storm drains, and rainfall variability can result in increased frequency, intensity, and duration of storm drain overflows. Inadequate source separation of domestic wastewater from rainwater and solid

---

[5]The issue of data—its generation, collection, aggregation and analysis—is extremely important in framing of policy-relevant dialogue, an issue we discuss in greater detail in Chaps. 3 and 4.

[6]See Kurian and Dietz (2013) for more information on this case study.

waste followed by necessary treatment can result in transport of contaminants into surface and groundwater sources that are important sources of water supply (Kurian and McCarney 2010). Over time, with deposition of solid waste, storm drains can silt up with consequences for public health because of mosquito breeding due to stagnant water, inundation of low-lying slum localities, and destruction of crops under peri-urban agriculture. However, on the flip side, when wastewater is better managed, significant economic benefits can be derived through reuse that advances multiple uses of environmental resources in sectors as varied as agriculture, poultry rearing, and homestead gardens (Jimenez and Asano 2008).

Double cropping and lower input costs for agriculture could be some of the direct benefits of wastewater collection and reuse (Rijsberman 2004). There could also be important economy-wide benefits of supporting freshwater swaps through use of treated domestic wastewater for agriculture. Source sustainability of urban water supply could be enhanced, agricultural productivity may increase and farm incomes can rise. However, climate-induced risks posed by variability in climatic, soil, and groundwater conditions can influence biophysical processes (*example, material flows*) and potential for cooperation thereby impacting upon realizations of hypothesized benefits.

This case study reports on the findings of a secondary review of 121 towns in India combined with a case study of a small town in the 0.2–0.5 million category. The study posed three key questions: (1) What are the economic benefits and costs of wastewater reuse in the context of hypothesized links to climate variability. (2) What is the role of local farming practices, market conditions, and crop variety in influencing wastewater reuse in agriculture? (3) What is the role of inter-governmental financing in influencing selection of technical adaptation options for collection, treatment, and disposal of wastewater?

3.2.2 The Idea of Risks for Wastewater Management in Urban Agriculture

Wastewater reuse for irrigated agriculture has often been viewed as a pathway to advance the nexus approach to management of environmental resources—water, soil, and waste. However, climate, market conditions, and technical options can seriously affect the translation of potential into actual economic benefits. A serious examination of inter-dependencies in resource use patterns cannot overlook sanitation challenges in peri-urban regions, public health risks, economic value of wastewater, and variations in climate. This case study based the assessment of wastewater reuse for peri-urban agriculture by considering off-site sanitation options (Kurian and Dietz 2013):

- Eco-san: Composting toilets with/without urine diversion;
- Settled sewerage: Involving sewer system receiving solids-free effluent from a septic tank, secondary wastewater treatment, and effluent reuse in aquaculture, agriculture, and horticulture; and

- Simplified sewerage: Systems receiving unsettled domestic wastewater. Simple junction boxes are used rather than manholes thereby reducing unit costs of systems.

We consider urbanization, notably growth of small towns in peri-urban regions, to be an important policy consideration, especially in developing countries and emerging economies. In light of rising population densities in peri-urban regions with adoption of simplified sewerage systems, a 20–50 % cost reduction may become possible when compared to conventional sewerage. Simplified sewerage offers benefits in the form of simpler technologies, lower unit costs, possibilities for wastewater reuse, scope for cost-recovery, and possibility to do away with on-site water supply for flushing of toilets. However, it is important to note that citizen participation combined with political support from mayors and public sector agencies is crucial for adoption of simplified sewerage technologies (Allen 2010).

Climate variability poses new challenges to advancing wastewater reuse schemes in peri-urban regions. Regional climate change scenarios predict major shifts in climate zones, although with much uncertainty about regional specificities (Kurian and McCarney 2010: 50). Where shifts are predicted from sub-humid to semi-arid conditions, particularly in areas with a dense population and intensive water demands, major problems can be expected (Dietz et al. 2004). However, what is more disturbing is the temporal aspect. All climates have an element of variability: between seasons, between night and day, between quiet and stormy conditions. Seasons are changing and variability is becoming more extreme the literature suggests. Periods of drought are feared to happen more often, but particularly periods of extreme rainfall and storms are likely to increase, and become more extreme, with a higher concentration of rainfall in fewer days (or hours). From a nexus perspective, these trends could imply the need for adaptation mechanisms to cope with lower predictability and more extremes. Adaptation could mean a shift to more water sources, from a wider geographical environment (*more inter-dependence*). What institutional and technical mechanisms are available to predict and respond actively to risks posed by climate-change?

There are a number of risks associated with considering reuse options for off-site sanitation in urban areas. Three main reuse risks include: (1) pathogen transfer through contamination of groundwater based water supply sources, (2) pathogen transfer through contamination of food chain-crop quality, and (3) pathogen transfer through disposal of untreated wastewater into rivers. The 2008 revised World Health Organization (WHO) guidelines identified indicators for environmental, soil, and geo-hydrological parameters that address various reuse risks from water source to waste disposal. An attempt was made to identify reuse possibilities together with associated risks for a range of agro-climatic contexts, including high or dry regions. The WHO guidelines identified parameters for monitoring water quality depending on type of use (water for drinking or irrigation) and by levels of crop resistance to effluent pollution (WHO 2008).

3.2.3 Making Trade-Offs Explicit to Promote Accountability for Wastewater Collection, Treatment and Disposal

Rationale for Wastewater Reuse in Indian Context

Land area devoted to wastewater farming in India is estimated between 6900 and 100,000 ha (Scott et al. 2000). Apart from reducing water scarcity, especially in drought prone areas, wastewater could be highly beneficial for agricultural purposes due to its high nutrient concentrations. This nutrient value with proper management can be transferred to crops and reduce the application of fertilizers. Revenues generated by farmers could then be used to treat wastewater to mitigate its negative health impacts. Given the magnitude of wastewater generated, the extent of area irrigated could be more than 1.2 million ha in all class I cities and more than 0.35 million ha in cities with populations between 2 and 5 hundred thousand. The total revenue generated with proper water pricing (*water + nutrient value*) is US$48 and US$14 million respectively, for the two categories of towns (Kurian and Dietz 2013: 51).

The health risks involved for humans and livestock and the returns from crops make it disadvantageous, especially in comparison to crops grown under open well or river/tank/canal irrigation. This case study revealed that wastewater irrigation is characterized by low investment, low yields, and lower market price. Due to the higher nutrient value of wastewater, farmers do not apply pre-sowing fertilizer, which results in savings of approximately US$10 per ha (Kurian and McCarney 2010: 55). On the other hand, average yield under wastewater irrigation is lower by about 7 quintals per acre. This factor coupled with a lower market price for paddy cultivated using wastewater results in gross returns that are lower by approximately US$200 per acre. However, our case study revealed that better management of wastewater can improve returns by approximately six times based on the ability to double crop and lower expenses on pesticides.

Implications for Policy: A Normative Focus

Public choice theory has emphasized that if sufficient autonomy is granted to local authorities, it may be possible to deal with the challenge of cost-recovery by mobilizing local finances and skills to address regional environmental challenges like wastewater pollution of rivers. However, on the contrary when central fiscal transfers do not allow for sufficient autonomy, it may be difficult to hold local authorities accountable for their revenue and expenditure practices (Veiga et al. 2015). For example, our case study revealed that when the local government at our study site was presented with technical options ranging from oxidation ponds (*with highest cost-benefit ratio*) to up flow anaerobic sewage blanket (*UASB—lowest cost-benefit ratio*), the authorities chose the UASB option. This was because central funds were readily available for its construction. Central transfers will only encourage dependence of sub-sovereign entities without emphasizing a search for cost-effective and efficient means of service delivery as long as central transfers are not tied to accomplishment of policy outcomes like connection of poorer households to a sustainable source of water supply or connection to a sewer network (World Bank 2006).

A search for sanitation options that promote pollution prevention rather than control, waste separation at source rather than end-of pipe treatment, and reuse of valuable nutrients rather than wasting by discharging into the environment can be supported by greater accountability in allocation of fiscal resources. In peri-urban regions, source separation of urine and rainwater are known to have the best prospects for improving water management (Wilsenach 2006). Urine separation could improve the efficiency of treatment processes, which would support the philosophy of "closing the loop" and recovery of nitrogen and phosphorous. Another futuristic option is to promote separation of feces where full separation occurs at the source through separate collection, handling, and treatment of yellow, grey, and black water. The wider applicability of this approach in developing countries remains to be seen. In institutional terms, the challenges that arise in developing countries could relate to the following:

- Identifying norms that would facilitate integration of water resource management from source to reuse by addressing issues of sectoral water allocation.
- Identifying norms for costing of water supply and sanitation interventions that would reflect the costs of separating waste at source.
- Identifying norms for billing of water supply and sanitation services, especially in contexts where multiple service providers from public and private sectors are involved (Salome 2010).

3.3   Case Study 3: Collective Action for Watershed Management in India[7]

3.3.1   Introduction

The Shiwalik hill forests comprising approximately 6.5 % of the land area of the North Indian state of Haryana perform an important ecological function of mitigating the effects of soil erosion. Open cattle grazing and fuel wood collection by local communities had endangered the soil erosion function of the Shiwalik forests. Forest degradation was manifested in the increasing rate of siltation of the Sukhna reservoir in the state capital, Chandigarh (Arya and Samra 1995). The reservoir was an important source of tourism revenue and analyses indicated that the source of the risk lay in rapid deforestation of the catchment areas located in proximity to the village of Sukhomajiri (Dhar 1994: 20).

---

[7]For more information on this case study, refer to Kurian et al. (2013).

A community-based forest management initiative that involved construction of water harvesting dams to arrest movement of silt was initiated. However, when villagers destroyed check dams that were constructed and continued to open-graze cattle in the forests, a substantive dialogue was initiated that resulted in usufruct sharing agreements between the forest department and local communities (Hill Resource Management Societies (HRMS)). The usufruct sharing agreements that initially covered fuel wood, fodder, and fiber grasses was later extended to include water from dams constructed by the Haryana Forest Department (HFD) with in-kind contribution of labor by the HRMS. The dams, by providing supplemental irrigation for the winter wheat crop, proved a powerful incentive in encouraging forest dependent communities to change their behavior from open grazing to stall feeding of livestock (Grewal et al. 1995). Multiple benefits emerged as a result of the Sukhomajiri model: increased fodder production on private fields and increased production of cattle dung used as cooking fuel in the area. As a result of increased production of fodder grass and cattle dung, a dramatic reduction in open cattle grazing and harvesting of fuel wood from forests was noticed, which in turn lead to forest regeneration and lower levels of soil erosion.

The case study that follows describes the importance of higher-order institutional rules in shaping environmental outcomes. More importantly, the case study is the first of its kind because it adopts a longitudinal approach to study the effect of institutional change on environmental management. By combining a survey approach with a comparative case study, the methodology adopted by the study emphasizes the importance of quantitative and qualitative approaches and its implications for benchmarking strategies that attempt to examine organizational performance.

3.3.2 Synergies Foster Collective Action Between Government and Community Organizations

Institutional Environment

Community-based forest management in the Shiwalik hills is not insulated from wider political economic trends occurring in Haryana in particular and India in general. Three trends are important. First, because of a secular decline in returns from pursuing a purely agricultural-based livelihood strategy, there has been a marked integration of farming populations in markets for non-farm labor and dairy products (Bhalla 1999; Varalakshmi 1993). Others have pointed out that given cultural norms in the region that prevent women from engaging in non-farm labor, women's workload in the domestic economy would have most likely risen to compensate for time spent away from the settlement by male members of households on account of their engagement in non-farm employment (Agrawal 1997). Second, imported paper and pulp have depressed demand for fiber grass sourced from the Shiwalik hill forests as a result of a larger trend of removal of import controls following the liberalization of the Indian economy (Kurian 1998). Finally, in response to growing profits of community-based forest management cooperatives fiscal mechanisms have become regressive (Kurian and Dietz 2007).

Public Auctions as Instrumental Basis of Co-production Contracts

The notion that public auctions could facilitate greater transparency gradually led the HFD to introduce auctions for sale of forest produce and rights to distribute water from earthen dams. Over time, a pattern evolved whereby contractors from outside the village bid higher than the HRMS at public auctions. Villagers preferred that contractors from within their village purchase rights to forest products and water distribution although no restrictions were placed on participation of outsiders. This preference is linked to their perception that a contractor drawn from within the village would be sensitive to individual household requirements for forest products and would likely accept instalment payments for services rendered. Further, local contractors were more likely to invest a share of the proceeds from sale of forest and water products in maintenance of the dam itself and indulge in gift giving in the form of donations to repair local schools and temples (Datta and Varalakshmi 1999: 117).

Infrastructure, Decision-Making and Accountability

The success of the Sukhomajiri project provided the HFD with the legitimacy it needed to go on a dam construction spree. Between 1985 and 1998, 45 water-harvesting dams were constructed in the Pinjore forest division. Lower level functionaries of the HFD exercised considerable discretion over decisions regarding design, operation, and maintenance of the earthen dams. In many cases, faulty design was overlooked to justify expensive repairs to parts of the dam that were washed away in the rains. The institutional framework encouraged collusion because villagers desired a share of the spoils of public investment, which took the form of provision of wage labor for dam construction and contract awards for specific tasks such as transport of distribution pipes from government warehouses to dam sites. However, in response to rising costs of infrastructure construction and evidence of rent-seeking behavior by field staff, the HFD resorted to public auctions whereby the highest bidder would be granted contracts for specific tasks.

3.4   Synergies for Community-Level Collective Action: The Role of Ecology, Power and History

In 2000, we undertook a survey of 28 community-based watershed management groups known as HRMS responsible for management of 45 earthen dams. Our survey revealed the complicity of the HFD in the faulty infrastructure design, which resulted in only 17 % of the dams functioning to their lifetime capacity of 20 years (Kurian and Dietz 2007). From a collective action perspective, we found that of the eight HRMS with functioning dams, those groups that exhibited successful cooperation were the ones characterized by heterogeneity in asset distribution and a higher interest in CPRs on account of: lower aggregate engagement in non-farm labor markets and limited access to private tube wells. To explore the relationship between group composition and collective action, we undertook a comparative case study of Bharauli and Thadion HRMS. The key findings of the study were as follows:

1. Land ownership, hierarchical social relations and labor market structure
   Bharauli is a village with 80 households while Thadion is a village with 50 households. Delicate power relations are evident in Bhrauli because it is a multi-caste village with a skewed pattern of land ownership. Hierarchical social relations are reflected in occupational specialization and segregation. In Thadion, a neighboring village, by contrast such caste-based patterns of hierarchy, social segregation or occupational specialization were absent. The structure of the labor market reinforces hierarchical social relations. For instance, landless households who work as hired labor on other people's fields or as domestic hands in the homes of the wealthy are not always paid in wages but in kind, such as food or insurance during times of natural disasters like floods. From a gender perspective, cultural norms that prevent participation of girls in the nonfarm labor market can make families with limited supply of male children highly reliant on farm production.

2. Differences in local ecology and location of farm plots
   Bharauli HRMS differed from Thadion HRMS with respect to local ecology (Wade 1988). Groundwater is relatively easy to find in Thadion while in Bharauli, the groundwater table is deep. The prohibitive groundwater drilling costs increased farmers' reliance on common pool resources in Bharauli where farmers also had access to kuhls for land irrigated by the dam. However, the kuhls run dry by early February and, if rains do not arrive by early March, the supply of the last round of supplemental irrigation for wheat depends on water from the earthen dam. Reliance on water from the dam becomes even more critical if rains fail altogether. Not surprisingly, we found that average land irrigated by earthen dams is the strongest explanation for understanding the degree of variance captured by household endowment scores.

3. Previous leadership experience
   Differences in caste or wealth status need not necessarily prevent groups from working together in a society that values democratic governance. For example, between 1995 and 2000, the water contractor (farmer Y) was the Sarpanch (chief) of the Bharauli local government. He had developed experience successfully leading a disparate group of people drawn from the different caste groups. Although caste rules prevented groups from inter-marriage or sharing of public space, they were fully capable of collaborating based on common but secular interests.

In 2008, we returned to our study site to examine how collective action had evolved over an eight-year period since our initial study. We found that as a result of several years of failure to promote collective action, the dam in Thadion had silted up and resource users had moved toward private water provisioning from tube wells. On the other hand, we found that collective action for dam management had continued in Bharauli. In the ensuing discussion, we adopt a longitudinal approach to examine the institutional basis of effective leadership and its role in delivering irrigation services.

Our longitudinal analysis of collective action for irrigation management points to three conditions of successful leadership:

1. Moral basis of power and authority

   An accurate description of power sources emerges when caste status is examined in the context of wealth differentials. Wealth ranking discussions revealed that irrigated land was considered a source of power. Interestingly, when we considered other factors such as land size or ownership of tractors, we were able to identify a different sub-set of individuals within the group. However, when we ranked households based on the composite endowment index that we developed, we were able to identify the water contractor as a powerful individual. He managed the largest acreage of irrigated land, owned a tractor, and possessed a small family, which implied fewer mouths to feed and potential to retain a large grain surplus that could be sold in local markets. Land sub-division at the time of marriage of male children has the potential to reduce levels of household wealth. However, the absence of male children in his family posed no secular threat to household wealth. Our focused group discussions revealed the contractor to be a benevolent patron despite his ability to control access to credit, land, and labor in the village.

   When crops failed due to pests or drought, the water contractor has traditionally been a source of credit at times. He also is known to provide loans for weddings, thereby cementing his position in the moral economy of the village. In the past, when loans were not repaid on time, the ownership of land that was pledged as collateral was transferred to the water contractor after allowing for a sufficient grace period. During harvest periods when family labor is insufficient to perform harvesting and threshing tasks, labor from landless households is hired. Laborers are not always paid in cash. Our longitudinal analysis revealed that eight years later the same conditions of in kind payments for labor contributions was retained.

2. Political factions

   In 1995, the water entrepreneur did not own the largest area of land in the village. That distinction belonged to another powerful person in the village, farmer X, who represents a powerful political faction. Political factions in the village are typically represented by members of the extended family of the patron (brothers, cousins) and clients (landless laborers). There are complex social norms that dictate behavior within factions. For example, an unwritten rule during a water auction is that members of a particular faction will not compete once one of their members has decided to place a bid. Farmer X dominated politics in the village prior to arrival of irrigation in the village and he served as liaison with local government. There was intense jostling among farmers to get irrigation to their fields. While allowing for constraints imposed by topography, when the pipes were finally laid, the water contractor had the largest acreage under irrigation in the village. Farmer X, with the larger family and more mouths to feed, was no longer in a position to exercise his power as he once used to do. Power had shifted gradually to a new political faction led by the water contractor. Incidentally, the two entrepreneurs who attempted a leadership role but failed in 2004, did not belong to either of the two political factions in the village.

3. Calculus of profit

Our revisit in 2008 revealed that de-silting of the dam pondage area was a pressing requirement. However, the important question that arose in this context was what type of contract form (*group provision vs. entrepreneur*) was better placed to bear the political risks associated with required tariff increases to facilitate adaptive environmental management. The combined effect of increasing group size (*facilitated by land fragmentation*) and lower tariffs (*shaped by emergence of countervailing forces within HRMS*) has potential to alter the "threshold of entrepreneur led collective action." The threshold is a function of: (1) marginal revenue derived from extending the irrigation network to accommodate new resource users who are paying a reduced tariff, and (2) availability of public subsidies that would enable the entrepreneur to retain the possibility of making a profit while charging poor consumers a lower tariff and undertaking de-silting of the dam pondage area.

3.4.1 The Usefulness of Transdisciplinary Methods

Our use of mixed methods advances transdisciplinary methods because: the composite index of interest and endowments based on food security assessments constitutes an improvement over single metric measures (*land and income*) of wealth distribution while allowing for benchmarking and comparisons over time; and forest vegetation analysis makes it possible to link dam management to condition of forest and soil resources. Such an approach is better placed to understand the institutional processes by which differential access to assets (*irrigated land, credit or labor*) at the level of individual households are translated into power (White 1989).

Power is understood here as the extent to which one could control the actions of others; from recourse to use of cultural symbols like gift-giving (Thapar 1994). Power is also exercised through patronage relations between landed and landless households that are embedded in labor-tying and informal credit arrangements that serve to insure groups against market and climate-based risks (Kozel and Parker 2003; Scott 1976).

4   Conclusions

At the beginning of this chapter, we pointed out that without a robust analytical framework, the science-policy divide will continue to support a disconnect between development and achievement of outcomes in terms of well-being and/or environmental sustainability. Based on the analysis undertaken of the three case studies in this chapter, we emphasize the importance of analyzing trade-offs and synergies involving different levels of government, private service providers, and communities. In this context, we underlined the principles of accountability, subsidiarity, and autonomy of decision-making. We therefore propose that the concept of co-provision can provide the basis for development of such an analytical framework that elaborates upon the following key elements (Kurian and Ardakanian 2015):

- Accountability in fiscal relations that facilitate identification of incentives for Operation and Maintenance (O&M) of infrastructure relating to water, soil, and waste services.
- Integrated analysis of *system* performance in terms of biophysical processes (e.g. *material flows*) or infrastructure operation (e.g. *dams or wastewater plants*) to cope with climate-induced risks posed by variability in climactic, soil, and groundwater conditions.
- Administrative culture and its influence on extent of discretion exercised by public officials in enforcement of rules relating to delivery of critical environmental services at different levels of government.
- Uncertainty in factor and product markets that influences incentives for cooperation in management of common pool resources.
- Contract forms that support the development of local leadership models to enforce rules for management of environmental resources.

Another major conclusion offered by analysis of three case studies relates to methodology. For the nexus approach to management of environmental resources to be advanced, it is important that transdisciplinary methods be used more widely. Combination of qualitative and quantitative methods, surveys and case studies are necessary to provide a robust analysis of the links between human and agency behavior, management regimes, and environmental outcomes. A major conclusion is that the relationship outlined above need not be a linear one (Scoones 1999). Complexity, heterogeneity, and variability characterize the relationship between biophysical and institutional domains when it comes to the nexus approach. The nexus approach could be advanced by adopting mixed methods and transdisciplinary approaches to research design, data collection, analysis, and presentation. Transdisciplinary approaches would entail finding innovative ways by which narrow disciplinary "silos" can be breached that would allow for social learning, trial and error, and incorporation of local insights in program planning and implementation. Transdisciplinary approaches would further support a rethink of the processes and structures that enable knowledge construction and dissemination with the potential to influence policy-making regimes.

The final conclusion this chapter offers is that with proper attention to evaluation design, important and generalizable principles can be gleaned from case studies of "success" and "failure" as they relate to management of water, soil, and waste resources. In Chap. 3, we discuss the role of data observatories and the potential role they can play in bridging the science-policy divide. We point out how the Nexus Observatory builds upon the principle of dispersed problem solving through access to a widespread network of project databases. Data observatories could potentially consolidate knowledge and support its translation into policy-relevant advice (Kurian and Meyer 2014). In Chap. 4, we demonstrate that the Nexus Observatory could support rapid scale up of nexus management approaches by facilitating the use of indices to advance evidence-based decision-making. Such a project would go a long way in demystifying concepts of trade-offs and synergies and provide a spatial and temporal context to application of nexus management tools to address challenges of environmental sustainability.

References

Agrawal, B. (1997). Gender, environment and poverty inter-links: Regional variations and temporal shifts in rural India, 1971–1991. *World Development, 25,* 23–52.

Allen, A. (2010). Neither rural nor urban: Service delivery options that work for the peri-urban poor. In M. Kurian & P. McCarney (Eds.), *Peri-urban water and sanitation services—Policy, planning and method* (pp. 27–61). Dordrecht: UNU-Springer.

Apgar, J. M., Argumedo, A., & Allen, W. (2009). Building transdisciplinarity for managing complexity: Lessons from indigenous practice. *International Journal of Interdisciplinary Social Sciences, 4*(5), 255–270.

Aree, J. Y., Meusch, E., Homsambath, K., & Feliciano, E. (2003, November 17–21). *Women, water and household food security in selected Laos PDR communities.* Chiang Mai, Empress Hotel.

Arya, S. L., & Samra, J. S. (1995). *Socio-economic implications and participatory appraisal of watershed management project at Bunga (Bulletin No. T-27/C-6).* Chandigarh: Central Soil and Water Conservation Research and Training Centre.

Bertea, M., (2005). *Transdisciplinarity and education: "The treasure within"—Towards a transdisciplinary evolution of education.* Paper presented at The Second World Congress of Transdisciplinarity, September 2005, Brazil. http://cetrans.com.br/artigos/Mircea_Bertea.pdf

Bhalla, S. (1999, January/April). Liberalization, rural labour markets and the mobilization of farm workers: The Haryana story in an all India context. *Journal of Peasant Studies, 26,* 62–88.

Biggs, S., & Smith, S. (2003). A paradox of learning in project cycle management and the role of organizational culture. *World Development, 31*(10), 1743–1757.

Dasgupta, P. (2001). *Human well-being and natural environment.* Oxford: Oxford University Press.

Dasgupta, S., Deichmann, U., Meisner, C., & Wheeler, D. (2005). Where is the poverty-environment nexus? Evidence from Cambodia. *Laos PDR and Vietnam. World Development, 33*(4), 617–638.

Datta, S., & Varalakshmi, V. (1999). Decentralization: An effective method of financial management at the grassroots. *Sustainable Development, 7,* 113–120.

De Rouw, A., Kydd, J., Morrison, J. & Poulton, C. (2003, December 15–23). Four farming systems: A comparative test for erosion, weeds and labour input in Luang Phabang District. *Juth Pakai, 1.*

Dhar, S. K. (1994). Rehabilitation of degraded tropical forest watersheds with people's participation. *Ambio, 23*, 72–84.

Dietz, A., Ruben, R., & Verhagen, J. (Eds.). (2004). *The impact of climate change on drylands.* Dordrecht: Kluwer.

Evrard, O., & Goudineau, Y. (2004). Planned resettlement, unexpected migration and cultural trauma in Laos. *Development and Change, 35*(5), 937–962.

Gibbons, M., et al. (1994). *The new production of knowledge—The dynamics of science and research in contemporary societies* (10th ed.). London: SAGE Publications.

Grewal, S., Samra, J., Mittal, S., & Agnihotri, Y. (1995). *Sukhomajiri concept of integrated watershed management (Bulletin No. T-26/C-5).* Chandigarh: Central Soil and Water Conservation Research and Training Institute.

Hirsch Hadorn, G., Bradley, D., Pohl, C., Rist, S., & Wiesmann, U. (2006). Implications of transdisciplinarity for sustainability research. *Ecological Economics, 60*, 119–128.

Jahn, T. (2008). Transdisciplinarity in the practice of research. In M. Bergmann & E. Schramm (Eds.), *Transdisziplinäre Forschung. Integrative Forschungsprozesse verstehen und bewerten* (pp. 21–37). Frankfurt: Campus Verlag.

Jimenez, B., & Asano, T. (Eds.). (2008). *Introduction.* The Hague: International Water Association.

Keohane, R., & Ostrom, E. (1995). Introduction. In R. Keohane & E. Ostrom (Eds.), *Local commons and global interdependence: Heterogeneity and cooperation in two domains.* Thousand Oaks, CA: Sage.

Kozel, V., & Parker, B. (2003). A profile and diagnostic of poverty in Uttar Pradesh. *Economic and Political Weekly*, 385–403.

Kurian, M. (1998). Issues in newsprint sector reform. *Public Enterprise, 16*, 127–138.

Kurian, M. (2010). Institutions and economic development: A framework for understanding water services. In M. Kurian & P. McCarney (Eds.), *Peri-urban water and sanitation services— Policy, planning and method* (pp. 1–26). Dordrecht: Springer.

Kurian, M., & Ardakanian, R. (Eds.). (2015). *Governing the nexus: Water, soil and waste resources considering global change.* Dordrecht: UNU-Springer.

Kurian, M., & Dietz, T. (2005). *How pro-poor are participatory watershed management projects? An Indian case study.* Research Report No. 92. Colombo: International Water Management Institute.

Kurian, M., & Dietz, T. (2007). *Hydro-logic: Poverty, heterogeneity and cooperation on the commons.* New Delhi: Macmillan.

Kurian, M., & Dietz, T. (2013). Leadership on the commons: Wealth distribution, co-provision and service delivery. *The Journal of Development Studies, 49*(11), 1532–1547.

Kurian, M., & Lestrelin, G. (2004). *Market integration, natural resource degradation and poverty: a comparative analysis of Thailand and Lao PDR.* Chiang Mai. Chiang Mai Hill Hotel, 5–9 September.

Kurian, M., & McCarney, P. (Eds.). (2010). *Peri-urban water and sanitation services: Policy, planning and method.* Dordrecht: Springer.

Kurian, M., & Meyer, K. (2014). *UNU-FLORES Nexus Observatory Flyer.* Dresden: UNU-FLORES. Retrieved from https://nexusobservatory.flores.unu.edu/

Kurian, M., Ratna Reddy, V., Dietz, T., Brdjanovic, D. (2013). Wastewater reuse for periurban agriculture—A viable option for adaptive water management?. *Sustainability Science, 8*(1), 47–59.

Kurian, M., & Turral, H. (2010). Information's role in adaptive groundwater management. In M. Kurian & P. McCarney (Eds.), *Peri-urban water and sanitation services—Policy, planning and method* (pp. 171–191). Dordrecht: Springer.

Lal, R. (2015). The nexus approach to managing water, soil and waste under changing climate and growing demands on natural resources. In M. Kurian & R. Ardakanian (Eds.), *Governing the nexus: Water, soil and waste resources considering global change.* Dordrecht: UNU-Springer.

Lang, D. J., et al. (2012). Transdisciplinary research in sustainability science: Practice, principles, and challenges. *Sustainability Science, 7*, 25–43.

Max-Neef, M. A. (2005). Foundations of transdisciplinarity. *Ecological Economics, 53*, 5–16.

MRC. (2003). *Mekong river basin profile: Towards the development of strategic research and development plans.* Vientiane: Mekong River Commission Secretariat.

North, D. (1990). *Institutions, institutional change and economic performance: Political economy of institutions and decisions.* New York: Cambridge University Press.

Ostrom, E. (1990). *Governing the commons: The evolution of institutions for collective action.* New York: Cambridge University Press.

Ostrom, E. (1996). Crossing the great divide: Co-production, synergy and development. *World Development, 24*(6), 1073–1087.

Ostrom, E. (2009). A general framework for analyzing sustainability of social-ecological systems. *Science, 325*(5939), 419–422.

Pohl, C. (2008). From science to policy through transdisciplinary research. *Environmental Science and Policy, 11*, 46–53.

Poteete, A., Janssen, M., & Ostrom, E. (2010). *Working together—Collective action, the commons and multiple methods in practice.* Princeton: Princeton University Press.

Reddy, V. R., Kurian, M., & Ardakanian, R. (2015). *Life-cycle cost approach for management of rnvironmental resources: A primer.* Dordrecht: UNU-Springer.

Rijsberman, F. (2004). Sanitation and access to clean water. In B. Lomborg (Ed.), *Global crisis, global solutions.* London: Cambridge University Press.

Roder, W. (1997). Slash and burn systems in transition: Challenges for agricultural development in the hills of Northern Laos. *Mountain Research and Development, 17*(1), 1–10.

Salome, A. (2010). Watewater management under the Dutch water boards - any lessons for developing countries? In M. Kurian & P. McCarney (Eds.), *Peri-urban water and sanitation services—Policy, planning and method* (pp. 111–131). Dordrecht: Springer.

Schreier, H., Kurian, M., & Ardakanian, R. (2014). *Integrated water resources management: A practical solution to address complexity by employing the nexus approach.* Working Paper—No. 2. Dresden: UNU-FLORES.

Scoones, I. (1999). New ecology and the social sciences: What prospects for a fruitful engagement? *Annual Review of Anthropology, 28*, 479–509.

Scott, C., Zarazua, J., & Levine, G. (2000). *Urban wastewater reuse for crop production in the water short Guanajuato River basin, Mexico.* Research Report No. 41. Colombo: International Water Management Institute.

Scott, C. A., Kurian, M., & Wescoat, J. L. (2015). The water-energy-food nexus: Enhancing adaptive capacity to complex global challenges. In M. Kurian & R. Ardakanian (Eds.), *Governing the nexus: Water, soil and waste resources considering global change* (pp. 15–38). Dordrecht: UNU-Springer.

Scott, J. C. (1976). *The moral economy of the peasant-rebellion and subsistence in south-east Asia.* New Haven: Yale University Press.

Scott, J. C. (1998). *Seeing like a state: How certain schemes to improve the human Condition have failed.* New Haven: Yale University Press.

Sen, A. K. (1999). *Development as freedom.* USA/UK: Oxford University Press.

State Planning Committee. (2000). *The households of Lao PDR: Social and expenditure indicators.* Vientiane: National Statistical Centre.

Thapar, R. (1994). *Cultural transaction in early India.* Oxford: Oxford University Press.

Turral, H. (1998). *Hydro-logic: Reform in water resources management in developing countries with major agricultural water use: Lessons for developing nations.* London: Overseas Development Institute.

United Nations. (2014). *Prototype global sustainable development report.* New York: United Nations Department of Economic and Social Affairs, Division for Sustainable Development.

UNU-FLORES. (2015). The nexus approach to environmental resources' management. Retrieved from https://flores.unu.edu/about-us/the-nexus-approach/

Varalakshmi, V. (1993). *Economics and goat and buffalo rearing - A case study from Haryana (Joint Forest Management Series, No. 4).* New Delhi: Tata Energy Research Institute.

Veiga, L. G., Kurian, M., & Ardakanian, R. (2015). *Intergovernmental fiscal relations: Questions of accountability and autonomy*. Dordrecht: UNU-Springer. (Springer Briefs in Environmental Science).

Wade, R. (1988). *Village republics: Economic conditions of collective action in south India*. New York: Cambridge University Press.

Wester, P., & Warner, J. (2002). River basin management reconsidered. In A. Turton & R. Henwood (Eds.), *Hydro-politics in the developing world: A southern African perspective* (pp. 61–71). Pretoria, South Africa: CIP, University of Pretoria.

White, B. (1989). Problems in the empirical analysis of agrarian differentiation. In G. Hart, A. Turton, & B. White (Eds.), *Agrarian transformations: Local processes and the state in South East Asia* (pp. 45–61). Los Angeles: California University Press.

WHO. (2008). *Guidelines for wastewater reuse in agriculture*. Geneva: World Health Organisation.

Wilsenach, J. (2006). *Treatment of source separated urine and its effect on wastewater systems*. Ph.D. Thesis, Delft University of Technology.

World Bank. (2006). *Fiscal decentralization in India*. Oxford: Oxford University Press.

Chapter 3
Political Decentralization and Public Services

Can Data Observatories Advance the Nexus Approach to Environmental Governance?

Abstract In this chapter, we draw upon two cases that employ results-based financing strategies to discuss elements of a decentralization framework that can potentially support enhancement of water and sanitation services. We also devote a part of the chapter to discuss the role that web-observatories can play in identifying generalizable principles based on analysis of case studies of "success" and "failure" as they relate to public services. A key argument we make in this chapter is that observatories, by consolidating knowledge and supporting its translation into policy-relevant advice, can go a long way in enhancing accountability and autonomy in revenue and expenditure decisions surrounding infrastructure construction, operation, and maintenance with potential to impact positively on the achievement of service delivery outcomes in particular and development goals in general.

Keywords Data · Monitoring · Environmental Resources · Risks · Governance · Public Services · Case Studies · Nexus Observatory · Index · Visualization · Benchmarking · Scenario analysis · Trade-offs

1 Introduction

The delivery of critical public services such as water supply, irrigation or wastewater treatment in developing countries and/or emerging economies suffers from fragmented approaches to planning and policy implementation (World Water Council 2015). Infrastructure construction, operation, and maintenance are critical nodes at which the fragmentation in planning and policy implementation is exacerbated (Kurian and Ardakanian 2015b). This is because revenue and expenditure decisions that relate to infrastructure are characterized by a lack of accountability and autonomy. As a result, the disconnect between development goals and achievement of outcomes and impact in terms of poverty reduction and

M. Kurian et al., *Resources, Services and Risks*,
SpringerBriefs in Environmental Science, DOI 10.1007/978-3-319-28706-5_3

environmental sustainability continues to persist. In the previous chapter, we pro-posed the use of a co-provision framework and argued that accountability may be enhanced if greater attention is paid to the following issues (Kurian and Dietz 2013):

- Fiscal relations that influence incentives for Operation and Maintenance (O&M) of infrastructure relating to water, soil, and waste services.
- Administrative culture that influences the extent of discretion exercised by public officials in enforcement of rules relating to delivery of critical environ-mental services at different levels of government.
- Contract forms that influence the development of local leadership models to enforce rules for management of environmental resources.

In the introduction to this Brief, we pointed out that far from being a linear process involving uptake of scientific advice, decision-making may entail having to "muddle through" based on important political trade-offs that may neither promote equity nor efficiency goals. While scientists may strive to achieve precision with their results, an effective bridge to the policy domain should strive to make trade-offs more explicit through use of transdisciplinary approaches.[1] It is worth re-emphasizing that this fundamental shift in perspective has several implications. First, it acknowledges the significance of decentralization (political, fiscal, and administrative) and its potential to affect decisions and development outcomes at scale. Second, it acknowledges that once trade-offs are made explicit, individuals and public agencies will be encouraged to design incentives that promote synergies that address common challenges such as water scarcity or pollution. Third, it is important to recognize that for solutions to emerge, data that is reliable, frequent, and sufficiently well disaggregated is important to ensure that decision makers can predict the scale and intensity of the policy challenge and bring to bear a propor-tionate amount of human and financial resources to realize the achievement of clearly verifiable development outcomes and impact.

In this chapter, we draw upon two cases that employ results-based financing strategies to discuss elements of a decentralization framework that can potentially support enhancement of water and sanitation services. We also devote a part of the chapter to discuss the role that web-observatories can play in identifying general-izable principles based on analysis of case studies of "success" and "failure" as they relate to public services. A key argument we make in this chapter is that obser-vatories, by consolidating knowledge and supporting its translation into policy-relevant advice, can go a long way in enhancing accountability and auton-omy in revenue and expenditure decisions surrounding infrastructure construction, operation, and maintenance with potential to impact positively on the achievement of public policy outcomes in particular and development goals in general.

---

[1]For an elaborate discussion on transdisciplinary approaches see Chap. 2 of this Brief.

2 Political Decentralization and Environmental Services: Key Considerations

The first chapter of the brief discussed three case studies that elaborated upon issues related to management of environmental resources—water, soil, and waste. The discussion considered the concept of the Poverty-Environment nexus and highlighted the importance of scale in determining whether environmental management actually impacted incidence of poverty. In this context, it is pertinent to point out that the delivery of services, such as irrigation, water supply or soil retention, are usually influenced by infrastructure considerations. When it comes to infrastructure, three factors (huge sunk costs, economies of density/scale and consumption) lead to politicization of pricing and operations in the water sector (Savedoff and Spiller 1999). Further, for a particular water infrastructure, increasing the number of connected households reduces the network's average operating costs. Consumers, especially in the developing world, normally associate water services as free goods. As a result, politicians can sometimes use the argument regarding pricing as an instrument for political mobilization (Kurian 2010).

The natural characteristic of water enables it to flow across multiple spatial scales. The provision of water services given the high capital costs involved and recurring maintenance expenditure necessitates coordinated management involving two or more political jurisdictions (United Nations Task Team on Habitat III 2015). The literature on the subject of coordination emphasizes three important principles of accountability, autonomy, and subsidiarity (Kurian and Ardakanian 2015b). The advancement of a number of the above principles highlights the importance of establishing a link between public expenditure on infrastructure construction and maintenance and revenues in terms of taxes and tariffs. Establishing a link between revenue, expenditure, and service delivery outcomes is predicated upon the availability of data that is disaggregated, reliable, and frequent. In this connection, assessments of "success" or "failure" of inter-governmental transfers in facilitating incremental improvements in service delivery would benefit from adopting a comparative framework (*based on locally identified indicators of quality, quantity, affordability or adequacy*), as opposed to an evaluation framework that is guided by transcendental policy goals (*examples include a 10 % community contribution toward the cost of operating a water system*). Real-time information flows that are multi-directional in nature (involving citizens and governments) can also go a long way in enhancing accountability of revenue and expenditure decisions at multiple levels of government (Kurian 2010).

In the wake of the Thatcher-Reagan revolution of the 1980s, privatization experiments involving public water utilities were undertaken. It was subsequently realized that privatization might not necessarily yield optimum results in terms of advancing efficiency or equity. In the developed world, experiments with forms of organization that lie between conventional public agencies and private companies were undertaken that included corporatization and performance-based organizations. Under the corporatization experiment, decision makers are removed from the influence of personnel, procurement, and budget restrictions that characterize public agencies. The focus instead is on identifying viable incentives. Performance-based organizations (PBO) on the other hand are government agencies that remain under public control, but in which agency officials are rewarded on the basis of performance.

In the case of developing countries, there is much to be said about the commitment to devolve revenue and expenditure decisions to local authorities (Kurian and Meyer 2015). A case in point being property taxes, which represent only 3–4 % of local revenues in developing countries, compared to 40–50 % in Australia, Canada, France, UK, and the US (United Nations Task Team on Habitat III 2015). More recently, several countries have begun experimenting with innovative schemes of municipal financing that are focused on improving service delivery. For example, Colombia has experimented with use of municipal funds as an instrument of sub-sovereign infrastructure finance, while cities in Mexico have begun encouraging local governments to improve their credit rating as a pathway to improve municipal finances and expand their financial resources. The case studies that follow discuss key lessons emerging from results-based financing initiatives in Asia and South America.

3 Results-Based Financing: Lessons on Planning and Implementation of Water Services Reform Projects

Two case studies of water services reform projects are presented in this section. They focus on results-based financing schemes that tie outcomes to funding in order to enhance the quality of service provision and strengthen accountability among the involved actors.[2] One of the most common results-based mechanisms used by international agencies, namely by the World Bank, on the delivery of basic infrastructure and social services to the poor is output-based aid (OBA). The OBA approach relies on a clear and realistic definition of objectives that can be measured and compared to the results accomplished. The first case study (UNU-FLORES 2015a) describes one of the first pilot projects, conducted by Global Partnership Output-Based Aid (GPOBA), designed to show that the OBA approach improves transparency and efficiency in the use of public resources, as well as the effectiveness in getting low-income families connected to water services networks. The second case study (UNU-FLORES 2015b) explores a pilot project, also conducted by GPOBA, to implement an OBA facility in Honduras to provide water and sanitation services to low-income families. The project tested whether it was possible to implement at the national level the type of subsidy scheme that GPOBA was using.

---

[2]See Veiga et al. (2015) for an overview of the recent experience in service delivery and financing models.

(a) *Case study 1*: *Output-Based Aid—Metro-Manila water supply improvement project*

Reducing the gap in the provision of safe water and sanitation services to low-income families continues to be an important challenge in many developing countries that traditional financing mechanisms do not seem to be able to address. Frequently, low-income households cannot afford to pay the connection fees required to access the services and have to resort to other alternatives, such as water tankers or other vendors, generally more expensive and of less quality. This case study focuses on one of the first pilot projects implemented by GPOBA to show that the OBA approach can be an effective way to help the poor, while enhancing the use of public funds in a transparent, accountable, and effective way. With OBA, the disbursement of funds depends on independent verification of the agreed results. Funds should be enough to guarantee that subsidies reach all intended beneficiaries, but final beneficiaries should have enough capacity to pay a proportion of the cost to access the service. This ensures a stronger sense of ownership of the project and a stronger commitment to pay the ongoing water and sanitation tariffs.

The study focuses on the water network in Metro-Manila in the Philippines. In the late 1990s, the government decided that the best way to improve water and sanitation services was to privatize the services. In 1997, two 25-year concession contracts, based on a geographic division of the region, were signed with two private firms. The contracts imposed several requirements in terms of the expansion of the coverage of water supply, sewerage, and sanitation services; the quality of the services; and the installation of public faucets in areas where individuals could not afford individual connection fees. One of the firms, the Manila Water Company (MWC) made important progress toward the fulfilment of the established objectives and was financially sound. However, the other firm faced financial problems and was unable to achieve the requirements. This led the government to re-tender the concession in 2007, and the new concessionaire showed good performance afterwards. The OBA pilot project was designed in 2007 and GPOBA decided to work only with the MWC given that the new concessionaire was initiating its activity.

In order to improve access to water in low-income areas, new pipelines were installed and the MWC worked closely with community leaders in the consultation and decision process. Although the programs improved the new consumers' access to water and reduced the associated costs, the connection fee was still too high for some households. Initially, to reduce connection fees, the MWC offered a shared meter option, where several households could share one mother meter and install individual sub-meters, with one household managing collection and remittance of payments to the concessionaire. Later, the MWC changed to a scheme of shared bills. Each community would manage a mini water distribution system, with a single account and a mother meter for the entire community. Each member of the community would have individual connections and sub-meters, but the community would be responsible for billing and collection for all its members. Even though prices were reduced, individual households were paying more than the community paid the concessionaire, and there was evidence of illegal tapping. To increase accuracy in meter reading and reduce illegal tapping, the MWC started installing several individual meters side-by-side inside protected structures (bank-arrangement), in accessible and strategic locations. However, there were still equity concerns, since some individual households free ride while the MWC was reluctant to disconnect the whole community. To overcome both problems at the same time, the MWC adopted a bank-arrangement scheme, combined with individual connections, with the support of an OBA scheme funded by GPOBA.

To ensure that the subsidies were provided to those that could not afford the services on their own, geographical targeting was used, since poor households are grouped in specific areas of Manila. Additionally, potential beneficiaries were asked to submit a certificate attesting that they were indigent. Given the sound operational and financial situation of the MWC, the GPOBA decided to adopt a single subsidy payment scheme after a single output was independently verified. The output was the provision of a working household connection that provided an acceptable service over a three-month period. Once the output was verified, GPOBA disbursed the unit subsidy multiplied by the number of verified connections to the MWC.

In sum, despite the clear effort of the government of the Philippines to enhance the efficiency and effectiveness of water service provision by selling concessions to private providers, the fee established by the service regulator proved to be too high for low-income households. The use of targeted results-based subsidies was crucial in ensuring more equity in water service access, and proved to be a transparent and efficient use of public funds to help low-income families becoming regular customers. Besides this direct effect, the project had other positive impacts, such as the reduction of water related diseases, reduction of household expenditure on water, and time savings for women per household.

This case study also demonstrated the need for a nexus approach to providing services to low-income areas. The project was successful in ensuring the connection of low-income families to the water supply network, but did not present a solution for how to dispose of the increased volume of wastewater. This unresolved wastewater problem prevented households from fully enjoying the benefits of the piped water, forcing them to use less than the desired volume of water because of the inappropriate sanitation facilities. A nexus approach by contrast would recognize the interdependence of resources and, that the management of one resource (access to water service) often creates challenges for other resources (wastewater management).

From the perspective of OBA, the Manila pilot project was deemed a successful case. It showed that OBA is an effective mechanism to improve equity of access to basic services, and promotes transparency and accountability in the use of public funds. The project provided relevant evidence that was later incorporated in the design of other projects implemented in several countries, using results-based financing mechanisms to improve the standard of living of the population. From a nexus perspective, however, the case study emphasized the need for approaches that acknowledge inter-dependence in use of environmental resources. The case also served as a basis for cross-fertilization of ideas based on implementation of service delivery reform projects in other parts of the world. This is why we argue in this Brief that data observatories have much to offer in terms of making the results of comparative case study analyses available in the form of generalizable principles that decision makers can draw upon to design, monitor, and evaluate developmental interventions covering water, soil, and waste resources.

(b) *Case study 2*: *Honduras Social Investment Fund: Using Output-Based Aid to provide water and sanitation services to low-income families*

Given the success of the OBA approach implemented by international agencies, as documented by the previous case study, pilot projects were carried out to test potential scale up mechanisms of having national governments implementing OBA schemes. The Manila case study highlighted the importance of a process of incremental learning that focused on modalities for targeting subsidies at the poor, independent verification of project outcomes, and design of concession contracts. The second case study from Honduras documents the challenges of scaling-up of institutional good practices to establish a national OBA facility in a country, to provide water and sanitation services to low-income families. From a nexus perspective, the Honduras case study has much to offer in terms of advancing transdisciplinary approaches to research, training and policy advocacy.

Honduras is one of the poorest countries in Central America. In a country where the service coverage for potable water and sewage was already deficient, rapid urbanization worsened the problem by creating peri-urban areas with no infrastructure to provide the services. Conscious of these problems, the government of Honduras adopted measures to improve the service provision through decentralization. In 2003, a new law was approved to restructure the water and sanitation sector. The national public monopoly was dismantled and local authorities were enabled to decide the service provision modality: public, private or mixed provision. Local service providers should be autonomous and financially viable, and operate under the oversight of a national regulator. Given the insufficiency of public resources to satisfy investment needs of water and sanitation in municipalities, national authorities were willing to explore alternative financing sources and schemes, creating an opportunity to introduce the OBA approach.

The GPOBA, on the other hand, had a good track record in the implementation of OBA to individual water and sanitation projects, and was open to experiment lending support in the setting up of an OBA facility in Honduras. They relied on the experience of a national entity (*Fondo Hondureño de Inversión Social*—FHIS) in the implementation of water and sanitation projects in the country. Data on local projects from FHIS was compared with data published in international reviews of project experience to develop unit costs to various outputs. The design of subsidy amounts for various elements of the project was guided by the following principles: 1. one of cost subsidies and not consumption subsidies; 2. disbursement of subsidies after independent verification of outputs; 3. Subprojects' tariffs should cover at least operation and maintenance costs; 4. pre-financing for private providers should be from internal cash generation or commercial loans; and 5. for public implementers (national and municipal institutions) pre-financing should be made available through loans provided by the OBA facility. The OBA facility was, therefore, housed within FHIS and supported by specialized consultants. It was responsible for screening, electing, prioritizing, and providing funding to subproject proposals fulfilling the FHIS and OBA eligibility criteria. The OBA facility also provided technical assistance in the design and implementation of the projects. The FHIS was fully responsible to the GPOBA for the compliance of the execution of the project.

The OBA facility started operating in 2008. After a first year of slow progress, several difficulties were identified in its implementation. It became clear that the OBA facility had not adequately consulted local authorities, GPOBA, potential contractors, NGOs, and community representatives. Meetings with the stakeholders were organized and adjustments were made to the project by FHIS and GPOBA allowing it to become a successful example of the first ever OBA facility to be implemented.

The implementation of this pilot project made it clear that, when designing future OBA facilities, it is important to: (1) Successfully integrate the facility in the overall financing framework of the sector; (2) Explain the innovative features and functionality of the OBA funding scheme to the stakeholders (local authorities, banks, contractors and service providers, and the population in general); (3) Assess the technical and analytical skills of the staff responsible for managing the facility, as well as their motivation to implement the new financing scheme; (4) Make sure that the implementers have the technical capacity to deliver the agreed upon results; (5) Adjust the geographical reach of the facility to the available subsidy resources in order to keep supervision and verification costs in check[3]; and finally, (6) Guarantee funding to implement a program at a scale adequate to the needs of the population, correctly synchronizing the OBA scheme with the overall financing architecture for water and sanitation projects.

The Honduras' national OBA facility pilot project was a source of inspiration for other national governments, such as that of Indonesia, to adopt results-based financing mechanisms. It proved that the implementation of a national OBA facility increased the effectiveness and transparency in the allocation of resources to subprojects aiming to reduce the access gap to water and sanitation services of low-income families. In nexus terms the Honduras case study highlights the importance of addressing issues of equity in addition to issues of technical and system efficiency. We have argued elsewhere that inter-governmental fiscal relations, notably, the role of taxes, transfers and tariffs are important considerations in advancing equity concerns in discussions on political decentralization and service delivery (Veiga et al. 2015).

---

[3]The transaction costs of the project were considered high relative to the size of the project, reducing its efficiency. This is not surprising since the project had a local focus. As Veiga et al. (2015) point out, notwithstanding the effectiveness of projects implemented according to results-based financing models, such as output-based aid, the high costs of data collection and project auditing are drawbacks that need to be tackled.

4 How Can Data Observatories Help Disseminate Institutional Good Practice?

4.1 Key Principles and Science-Policy Domain Goals

Case studies like the ones discussed in the previous section provide rich anecdotal evidence and descriptive detail for analysis. However, some important steps could support robust analysis of cases with the objective of identifying generalizable principles. These steps include creation of databases, design of databases, design of data collection protocols, data analysis procedures, and use of sampling criteria to specify units and levels of research. In many instances, several agencies and individuals may have already begun work on various dimensions of this challenge. In the case of environmental resources, such as water, soil, and waste, a number of UN agencies, government research institutes in the developing and developed world, and resource users have access to data in disparate forms (see Table 1). Data observatories have the potential to aggregate data from several sources and by using different medium *(for example, paper-based versus mobile)* to perform three functions that are crucial from the point of view of supporting evidence-based decision-making: data classification, knowledge consolidation, and knowledge translation.

The idea of a data observatory is founded on a few core principles as outlined below (Hall and Tiropanis 2012).

1. Access to distributed repositories of data, open data, online social network data, and web archive (Hall and Tiropanis 2012).
2. Harmonized access to distributed repositories of visual/analytical tools to support a variety of quantitative and qualitative (transdisciplinary[4]) research methods that are inter-operable with either published or private datasets.
3. Shared methodologies for facilitating the harvesting of additional data sources and the development of novel analytical methods and visualization tools to address societal challenges and to promote innovation.
4. A forum for discussion about an ethics framework on the archiving and processing of web data and relevant policies.
5. A data-licensing framework for archived data and the results of processing those data.

It can be argued that data observatories, if effectively managed, can contribute toward bridging the science-policy divide. To achieve this goal successfully observatories may strive to realize the goals as given in Table 2 along both science and policy domains (Kurian and Ardakanian 2015c).

---

[4]See Chap. 2 of this Brief.

Table 1 Three levels of engagement

Data (classification)	Knowledge (consolidation)	Information (conveyance)
UN agencies	Scale/boundary conditions/feedback loops	Trade-offs/synergies/resource optimisation
Member states	Scale/boundary conditions/feedback loops	Trade-offs/synergies/resource optimisation
Private data sets	Scale/boundary conditions/feedback loops	Trade-offs/synergies/resource optimisation

Table 2 Policy and science domain goals from the point of view of environmental resources, services, and risks

	Domain	Goals
From the point of view of environmental resources, services, and risks	Policy	1. Knowledge transfer through regional consultations and international conferences 2. Field testing new approaches to planning and management 3. Identify policy/program management triggers based on data visualization 4. Incubation of policy-relevant research questions based on proposal writing workshops emerging in the wake of regional consultations 5. Dissemination of good practice guidelines through publication of policy briefs
	Science	1. Specification of boundary conditions to heighten the applicability of research outputs 2. Specification of scale conditions to determine the applicability of research outputs 3. Identification of nexus intersections through examination of nodes in the biophysical, institutional, and socio-economic domains that influence the management of environmental resources, services, and associated risks 4. Identification of nexus interactions through examination of biophysical and institutional processes that impact upon the management of environmental resources, services, and associated risks 5. Identification of feedback loops that transmit the effects of policy/program interventions on human behavior and their consequences for the management of environmental resources, services, and associated risks

For a more detailed description, see Kurian and Ardakanian (2015a); especially pp. 225–229

4.2 From Decision Support Systems to Web-Observatories

The importance of availability of and accessibility to relevant, up-to-date, and reliable data and information that clarifies trade-offs and synergies have already been highlighted in this Brief. It was also demonstrated that this includes the classification of data as well as the consolidation, translation, and transfer of knowledge[5] (Kurian and Meyer 2014). Above all, the digital revolution is opening new cost-effective, easily attainable opportunities for knowledge management, analysis, and transfer. To support complex decision-making and problem solving, decision support systems emerged, which have been widely used in environmental risk management and covered in the literature.[6] Decision support systems in many instances are supported by technology that typically comprises tools that allow for easy management of data and knowledge, functionalities for modelling, and an interface that is accessible, interactive and easy to navigate (Shim et al. 2002). Through these components, decision makers, at the appropriate level, can approach problems from a more comprehensive perspective that draws on evidence and information that may lead to better choices with regard to management of environmental resources.

Bui points out that the use of decision support systems promotes "a changing consciousness about environmental responsibility" that will result in better informed decision-making, but cautions that "success or failure of sustainable development depends more on political and managerial leadership than on advanced technology" (Bui 2000: 3–4). Highlighting the reliance on individual decision makers underscores the limits of decision support systems, which need to be overcome, while offering great potential for producing sustainable outcomes. Therefore, governance processes and consideration of alternative options, such as those described in the case studies above, form an integral part of environmental planning and management as well as in the delivery of related services (e.g. water and sanitation).

Another weakness of many decision support systems related to sustainable development issues, is their specialized, local, and thematic problem focus (e.g. at village or watershed level) (Bui 2000), which does not promote cross-fertilization across regions. Additionally, these specified boundary and scale conditions do not necessarily account for political and/or institutional parameters, which define the spatial and temporal competencies of decision makers.

Apart from decision support systems on specific issues, such as water resources management or environmental impact assessment (Kersten and Lo 2000), a new

---

[5]Bui (2000) provides a more detailed, nevertheless not comprehensive, list for the definition of decision-making tasks: "comprehensive classification of problem types, information requirements, decision making procedures, selection of decision makers involved in the process and … the spatial and temporal impacts of sustainable development decisions."

[6]For an assessment of various Decision Support Systems, see Kersten et al. (2000).

trend is emerging, the use of web-based observatories. These observatories rest on the premise that they allow for the linking of various data sources. Doing so enables the integration of data as well as the closing of data gaps, so long as the reliability and quality of data can be guaranteed (see Table 1) (Fundulaki and Auer 2014; Terry et al. 2014). Links and interconnections between already available data will contribute to a "web of data" that is greater than the sum of its parts. Bundled into this "web of data," observatories will not only allow for greater accessibility to data and databases relevant to a particular theme, but also engender a systematic, time and resource efficient resolution to identify gaps and overlaps, as well as possible discrepancies between various sources of the same or similar data (Kyzirakos et al. 2014).[7] It also increases the frequency of available information, allowing for the use of near real-time data. Coupled with a mix of visual and analytical tools for quantitative and qualitative research, evidence-based decision-making can be strengthened and promoted. In the context of the envisaged Nexus Observatory, this is realized by focusing on the nexus between water, soil, and waste, thereby, making explicit the inherent synergies, trade-offs, and feedback loops across sectors (Kurian and Ardakanian 2015c; Kurian and Meyer 2014).

Considering the issue of overcoming data gaps further, the efforts surrounding the establishment of monitoring frameworks for Sustainable Development Goals (SDGs) targets is notable (GEMI 2015). Novel data collection approaches, examining the role of new technologies, should be considered to close such gaps. A useful entry point for supplementing often-incomplete data and information is the use of earth observation systems. These include satellite imagery, remote sensing, geoinformation systems (GIS), and in situ data collection that complement use of big data (Independent Expert Advisory Group on a Data Revolution for Sustainable Development 2014).[8] Web-based observatories have the potential to assist with the integration of earth observation data with other sources of information, such as surveys, legal documents, local registries, economic data, private data (see Table 1) etc. (United Nations 2014). Additionally, where it is viable to make available near real-time, high quality and reliable data that can be analyzed quickly, it would be possible to manage environmental risks, such as floods and droughts.

A key hypothesis we explore in the next chapter of this Brief is that improved accountability and autonomy in fiscal decision-making can lead to better management of environmental risks and outcomes.[9] Such an integrated approach for progress monitoring in interconnected sectors and clusters, where data harmonization has occurred and comparable standards have been established, will lead to

---

[7]This is to some extent comparable to astronomical observatories, where professional and hobby astronomers are encouraged to add any discoveries to an interactive web-based planetary observation system. Due to the increased number of actors monitoring different parts of the hemisphere, the scientific results can be multiplied enormously.

[8]For additional information, please refer to the United Nations Global Pulse initiative on big data, available at http://www.unglobalpulse.org/.

[9]See Chap. 4 of this Brief.

better results, generate political buy-in, and direct investment (human and financial resources) toward integrated and transdisciplinary development research, methods, and programs that advance sustainable development in general and equity goals more specifically.

4.3 The Appeal and Benefits of Using Web-Observatories

The academic discourse on web-observatories is still limited, focusing mainly on the issue of linking data through open data applications (ERCIM News 2014). The way in which the value of data observatories informing sustainable development discourse is presented in the present Brief is at the cutting edge of nexus research and implementation. A thorough analysis of nexus cases, such as those described above, offers greater insight into inherent complexities in varying situations, presents alternative options for arriving at sustainable solutions and permits comparisons between cases that may help identify research/implementation gaps (e.g. providing water connections, but overlooking the interconnections to wastewater and sanitation provisions) or generate generalizable, scalable principles qualified by regional or local specificities.

It comes as no surprise that comprehensive observatories that offer a holistic point of data and information for decision-making are gaining in relevance. This is possible due to advances in Information and Communication Technology (ICT), as well as the ever-increasing realization that we are living in an interconnected, interdependent world, which requires real world problems to be addressed in an integrated manner. Post-2015 development agenda debates are also supportive of the above analysis, emphasizing nexus approaches that focus on solving complex sustainable development problems taking into account economic, social, and environmental factors (United Nations 2014, 2015). As discussed in Chap. 2 of this Brief, employing transdisciplinary methods is a first step toward development of a more complete and comprehensive evidence-base that can effectively engage with nexus challenges.

The move toward web-observatories, in particular, in the field of sustainable development offers a number of opportunities. By their nature, as indicated previously, observatories have the potential to provide more comprehensive assessments, aggregate information to overcome fragmentation, allow working across disciplines, bridge the science-policy divide, and illuminate synergies and trade-offs, involving economic, social and environmental processes, and institutional structures. A web-observatory, as envisaged for the Nexus Observatory, will establish cross-sectoral assessments and evaluations of progress that highlight such interlinkages, synergies, and trade-offs as they apply to the nexus of water, soil, and waste. An approximation of science and policy can, thus, occur by taking advantage of scientific research on the nexus approach, multi-stakeholder engagement, and advances in ICT.[10]

As highlighted above, observatories, in contrast to decision support systems, generally address a number of issues, which belong to a particular theme. One example of such an observatory is the UN-Habitat Global Urban Observatory (GUO), which places urbanization processes and considerations at the center of enquiry (UN-Habitat 2012). The design of an observatory would then allow for a more holistic assessment with regard to these overarching themes. GUO was called into existence to assist with the monitoring of the implementation of the Habitat Agenda and target 11 of the Millennium Development Goals (MDG),[11] hence, measuring the progress of the state of urban development. GUO mainly serves as a database of urban statistics and indicators that are stored, presented, and analyzed through presentation tools. The data and information gathered [primarily through Geographical Information Systems (GIS)] as part of the monitoring activities contributes to a Global Urban Indicators Database, monitoring of urban inequities, and an Urban Info Database, covering a number of topics related to the urban context (e.g. housing, education, crime). Additionally, GUO provides evidence on urban development for related reports, including those on the MDGs (UN-Habitat 2012).

It is not apparent from the GUO platform, whether any topics or issues that are being monitored are analyzed in an integrated manner in order to promote sustainable resource management solutions (e.g. the role of local governments, decentralization or options for service delivery). Furthermore, whereas dissemination of data and information of progress and potential shortfalls in meeting internationally agreed targets are communicated through reports, the inclusion of an interface in which data or knowledge gaps can be identified and overcome using the web-observatory could enhance progress, development, and implementation even further. Hence, the science-policy divide remains insufficiently addressed. However, the significance of GUO's contribution to data and information aggregation, monitoring efforts, and achievements in multi-stakeholder engagement (mainly ministries at the national level) must not be overlooked.

Nonetheless, it is clear that a web-observatory with a primary focus on monitoring coordination of the use of indicators offers limited advantages from a nexus perspective. Even at present, a number of important aspects have received only partial attention. These relate to the identification of emerging issues, provision of substantive capacity building, contributions to/of scientific research (including research funding) that can in turn inform policy,[12] and alternative management options that impact service delivery and fiscal systems. Case studies, such as the

---

[10]This is also in line with proposed SDG Goal 17, which highlights technology, capacity building, multi-stakeholder partnerships, and data, monitoring and accountability among other issues.

[11]MDG Target 11: Have achieved by 2020, a significant improvement in the lives of at least 100 million slum dwellers, http://www.unmillenniumproject.org/goals/gti.htm.

[12]As mentioned above, this also relates to insufficient cross-fertilization, where integration across sectors and disciplines do not occur and comparative and/or collaborative research at regional

ones discussed in this chapter, can shed light onto these processes. Observatories in our view have a very powerful function in transferring the insights gleaned from case studies in a form that can be understood and used by decision makers.

Similar to UN-Habitat, the World Health Organization (WHO) has developed a Global Health Observatory and is currently in the process of developing an additional Global Observatory on Health Research and Development (R&D) (WHO 2015; Terry et al. 2014). As observatories, including GUO, created and regulated by international organizations with specific mandates and guided by internationally negotiated goals and targets, such as the Millennium Development Goals and from 2016 the Sustainable Development Goals, it is not surprising that the aggregation of relevant data, monitoring against international targets, and measuring progress over time are main priorities. However, with reference to the WHO Global Observatory on Health R&D, there seems to be an understanding that the mapping of research and associated funding may be sufficient to arrive at an understanding of data, knowledge, and capacity gaps. It focuses on "How to finance research and development where normal market forces are absent" (Terry et al. 2014: 1302). In the same way as the aforementioned case studies on results-based financing engage with questions of public services provision and financing, so too does the Global Observatory on Health R&D address questions of utilization of limited resources in line with policy priorities in the public health domain (Terry et al. 2014). This leads us to conclude that observatories have the potential to improve coordination in and across both the science and policy domains, in particular, with a view to identifying generalizable principles.

In the case of environmental resources management and governance, a web-observatory has the potential to contribute considerably to advancing research, discourse, and implementation on the nexus approach to the management of water, soil, and waste. The holistic architecture of such an observatory allows for a more realistic grasp of real-world problems, while promoting the interface between science and policy. The Nexus Observatory at UNU-FLORES is an ambitious undertaking as it endeavors to integrate a vast amount of knowledge, methodologies, data, tools, capacity building programs, etc. related to the nexus of water, soil, and waste into one web-based system (Kurian and Meyer 2014). In contrast to the separation of tasks of observatories, like those by WHO, UNU-FLORES endeavors to create an integrated experience that moves beyond systems that are focused only on making available data, monitoring or research classification and instead aims to create an interface that allows for research to inform capacity development and policy processes and vice versa.

While the conceptualization of the Nexus Observatory with its various components is a demanding task and research is ongoing, development of its numerous elements will have to take place gradually, continuously improving based on

---

(Footnote 12 continued)

(e.g. East Africa) or cross-regional (e.g. Africa–Asia) do not occur. See also Chap. 2 of this Brief and regional consortium formation (Kurian and Meyer 2014).

experience. Due to technical challenges and the state-of-the-art nature of the Nexus Observatory, sustainability of the platform depends on durable commitments, buy-in, and support. It follows that the full set of functionalities can only become operational over time, depending on resources (both human and financial), the willingness of partners to collaborate over an extended period of time, to provide relevant data, information and other inputs, and work toward a common goal. The consortium approach that we will elaborate on in Chap. 4, has proved effective in generating trust and political buy-in based on agreements for data sharing and collaborative research that can inform investments in sustainability research. The cooperation agreements that support the consortium approach are premised on demands and priorities determined by Member States and articulated at regional consultations that were organized in Africa and Asia.[13]

Compared to other observatories, such as the ones described above, the Nexus Observatory is characterized by its state-of-the-art links between scientific research and implementation, while allowing for application of transdisciplinary methods and approaches through acknowledgement of sites of knowledge and design of hybrid methodologies for data collection and analysis. Additionally, taking a nexus perspective and particularly including waste as a resource offers a new dimension that has not previously been addressed. It goes beyond thematic categorization and approaches sustainable development research in clusters of interconnected factors (Kurian and Ardakanian 2015a). A web-observatory, as defined in this Brief, therefore, through comprehensive and holistic analysis, elucidates intersections and interactions[14] (e.g. by analyzing case studies, such as those presented here) and provides an evidence-base for better decision-making.

5 Conclusions

The nexus approach has emphasized the management of environmental resources—water, soil, and waste. The biophysical domain has been highlighted in discussions on the nexus approach so far. The research community has illuminated issues of inter-dependence and interconnections among "compartments" and put a spotlight on "fluxes" and "flows." These are all-important parts of the discussion, but in our view, demand for elements of the nexus approach will be driven primarily by its applications in practice. Decision makers work in an environment where they have to make continuous and sometimes quick decisions with regard to allocation of scarce budgetary and human resources. Those decisions are not driven purely by the

---

[13]A regional consultation is characterized by the participation of at least five states represented by researchers and/or decision makers (mostly ministerial level).

[14]Intersection and interaction refer to factors that define the scope and relevance of the nexus approach as well as vertical and horizontal impacts and structures of feedback loops affecting the management of environmental resources respectively. For a detailed account see Chap. 2 of Kurian and Ardakanian (2015a).

need to promote "resource" conservation but in many instances by the need to "sustain" delivery of critical public services such as irrigation, water supply or wastewater treatment. This takes us back to the question we posed in the introduction to this Brief that relates to why statistical significance of research results does not always coincide with political action.

The literature on political and administrative decentralization emphasizes the importance of revenue and expenditure considerations that guide decision-making when it comes to public services. Analysis of accountability, administrative culture, and contract models involving public-private partnerships assume great importance when it comes to enhancing the delivery of public services. In this context, infrastructure considerations are important in framing discussions on decentralization. Economies of scale, population density and sunk costs of infrastructure and distribution of benefits and costs arising from operating and maintaining infrastructure assume extreme importance. This chapter argued that a nexus approach must move beyond an exclusive focus on environmental resources to one that engages with challenges of balancing "efficiency" and "equity" goals that considerations of infrastructure impose upon decision-making structures and processes. For this purpose, improved data and its analysis as well as translation of knowledge into evidence that can be used by decision makers is an issue of critical importance.

This chapter offers a refreshing new perspective on how we may bridge the gap between a conceptual focus on resources and services in discussions on the nexus approach. We employed two case studies on results-based financing to highlight the links between technology choice and fiscal systems as they play an important role in mediating the delivery of water and wastewater services in the Philippines and Honduras respectively. The Philippines case study discussed issues relating to independent verification of project outcomes, contract models that engage the private sector and design of subsidy schemes that target poor consumers. The Honduras case study on the other hand highlighted the complicated political and administrative process that very often determines the success of scaling up development interventions. The role of international donors, capacity development, and data collection and analysis were discussed in this context.

Perhaps, the most important feature of this chapter is the section where we discussed the role of observatories in providing a context for analysis of "success" and "failure" of individual interventions. We argue that data observatories can play a powerful role in bridging the gap between science and policy by providing a seamless interface between big data applications, capacity development, and policy engagement. Moreover, data that has been properly classified and knowledge that is well organized by theme or regional priorities can lead to generation of generalizable principles that can guide decision-making. To support our arguments, we examined the experience with two ongoing initiatives that relate to the use of observatories within the UN system. We discussed the lessons we can learn from existing initiatives and asked how we may improve upon them to devise a Nexus Observatory that informs design and monitoring of interventions as they relate to management of environmental resources and services.

In the next and final chapter, we take the discussion on observatories forward by examining the application of a nexus index in informing decisions relating to management of risks as they apply to environmental services. For that discussion, we focus on two specific challenges that have been identified by the SDGs and African Member States as important policy priorities; notably droughts and floods.

References

Bui, T. X. (2000). Decision support systems for sustainable development: An overview. In G. E. Kersten, Z. Mikolajuk and A. Gar-On Yeh (Eds.), *Decision support systems for sustainable development: A resource book of methods and application*. New York: Springer.

ERCIM News. (2014). Special theme: Linked open data. *ERCIM*.

Fundulaki, I., & Auer, S. (2014). Linked open data. *ERCIM, 96*, 8–9.

GEMI. (2015). Monitoring wastewater, water quality and water resources management: options for indicators and monitoring mechanisms for the post-2015 period. *A discussion paper for GEMI*. Global Expanded Water Monitoring Initiative, 1st stakeholder consultation, Geneva, 29 and 30 January 2015. Retrieved from http://www.unwater.org/fileadmin/user_upload/unwater_new/docs/Topics/SDG/Discussion_Paper_Gemi_Geneva_Meeting_29-30Jan2015_FINAL2015-04-27.pdf

Hall, W., & Tiropanis, T. (2012). Web evolution and web science. *Computer Networks, 56*, 3859–3865.

Independent Expert Advisory Group on a Data Revolution for Sustainable Development. (2014). *A world that counts: Mobilising the data revolution for sustainable development*. New York: Data Revolution Group.

Kersten, G. E., & Lo, G. (2000). DSS application areas. In: G. E. Kersten, Z. Mikolajuk and A. Gar-On Yeh (Eds.), *Decision support systems for sustainable development: A resource book of methods and applications* (pp. 391–407). New York: Springer.

Kersten, G. E., Mikolajuk, Z., & Gar-On Yeh, A. (2000). *Decision support systems for sustainable development: A resource book of methods and applications*. New York: Springer.

Kurian, M. (2010). Institutions and economic development—a framework for understanding water services. In M. Kurian & P. McCarney (Eds.), *Peri-urban water and sanitation services—policy, planning and method* (pp. 1–26). Dordrecht: Springer.

Kurian, M., & Ardakanian, R. (2015a). *Governing the nexus—water, soil and waste resources considering global change. Springer Briefs in Environmental Science*. Dordrecht: UNU-Springer.

Kurian, M., & Ardakanian, R. (2015b). The nexus approach to governance of environmental resources considering global change. In M. Kurian & R. Ardakanian (Eds.), *Governing the nexus: Water, soil and waste resources considering global change* (pp. 3–13). Dordrecht: UNU-Springer.

Kurian, M., & Ardakanian, R. (2015c). Policy is policy and science is science: Shall the twain ever meet? In M. Kurian & R. Ardakanian (Eds.), *Governing the nexus: Water, soil and waste resources considering global change* (pp. 219–230). Dordrecht: UNU-Springer.

Kurian, M., & Dietz, T. (2013). Leadership on the commons: Wealth distribution, co-provision and service delivery. *The Journal of Development Studies, 49*(11), 1532–1547.

Kurian, M., & Meyer, K. (2014). *UNU-FLORES nexus observatory flyer*. Dresden: UNU-FLORES.

Kurian, M., & Meyer, K. (2015). *The UNU-FLORES nexus observatory and the post-2015 monitoring agenda*. Retrieved from https://sustainabledevelopment.un.org/content/documents/6614131-Kurian-The%20UNU-FLORES%20Nexus%20Observatory%20and%20the%20Post-%202015%20Monitoring%20Agenda.pdf

Kyzirakos, K., Manegold, S., Nikolaou, C., & Koubarakis, M. (2014). Building virtual earth observatories using scientific database and semantic web technologies. *ERCIM, 96*, 10–11.

Savedoff, W., & Spiller, P. (1999). *Spilled water: Institutional commitment in the provision of water services*. Washington, DC: Inter-American Development Bank.

Shim, J. P., et al. (2002). Past, present, and future of decision support technology. *Decision Support Systems, 33*, 111–126.

Terry, R. F., Salm, J. F, Jr, Nannei, C., & Dye, C. (2014). Creating a global observatory for health R&D. *Science, 345*(6202), 1302–1304.

United Nations. (2014). *Prototype global sustainable development report. Online unedited edition, issued 1 July 2014*. New York: United Nations Department of Economic and Social Affairs, Division for Sustainable Development.

United Nations. (2015). *Global sustainable development report: 2015 edition, advanced unedited version*. New York: United Nations Department of Economic and Social Affairs, Division for Sustainable Development.

UNU-FLORES. (2015a). *Output-based aid—Metro Manila water supply improvement project*. Dresden: UNU-FLORES Nexus Observatory.

UNU-FLORES. (2015b). *Honduras social investment fund: Using output-based aid top provide water and sanitation services to low income families*. Dresden: UNU-FLORES Nexus Observatory.

UN-Habitat. (2012). *Global Urban Observatory (GUO)*. Retrieved from http://unhabitat.org/urban-knowledge/global-urban-observatory-guo/

United Nations Task Team on Habitat III. (2015). *Habitat III issues papers: 7—Municipal finance*. New York: United Nations Task Team on Habitat III.

Veiga, L. G., Kurian, M., & Ardakanian, R. (2015). *Intergovernmental fiscal relations—questions of accountability and autonomy. Springer Briefs in Environmental Science*. Dordrecht: UNU-Springer.

World Health Organization (WHO). (2015). *Global Health Observatory (GHO) data*. Retrieved from http://www.who.int/gho/about/en/

World Water Council. (2015). *Integrated water resource management: A new way forward*. Discussion paper, 7th World Water Forum.

Chapter 4
Disaster Risk, Political Decentralization and the Use of Indices for Evidence-Based Decision Making

Applications of a Nexus Observatory

Abstract Disaster risk governance is a function of institutions at multiple levels of government to predict and effectively respond to threats posed by global changes such as urbanization, climate and demographic change. An important determinant of government's effectiveness in responding to environmental threats could be levels of administrative and political decentralization. In Chap. 3, we discussed two case studies and examined how results-based financing (RBF) models could support a strategy for the delivery of water and sanitation services. In this chapter, we outline tentative research propositions to discuss the role of decentralization in predicting and responding to disaster risk. We also outline the applications of a Nexus Observatory by discussing the role of the following tools: nexus index, data visualization, scenario analysis, and benchmarking. We argue that the above tools could support the use of RBF approaches that strengthen accountability in revenue and expenditure decisions of local authorities.

Keywords Data · Monitoring · Environmental resources · Risks · Governance · Public services · Case studies · Nexus observatory · Index · Visualization · Benchmarking · Scenario analysis · Trade-offs

1 Introduction

According to the United Nations Global Assessment Report on Disaster Risk Reduction (UNISDR 2013), the worst in disaster risk is yet to come. The report warns that, given the population growth, rapid urbanization in hazard-exposed countries, and investments that do not seriously take disaster risk into account, potential future losses are enormous. These losses may be evident through their effects on infrastructure. For example, prolonged drought may result in "sunk costs" in water infrastructure being under-utilized. Further, in the case of certain water users, such as in the case of irrigation, prolonged drought may result in non-payment of tariffs that compromises the financial sustainability of the scheme. In the case of wastewater infrastructure, financial sustainability is critical to ensuring

M. Kurian et al., *Resources, Services and Risks*,
SpringerBriefs in Environmental Science, DOI 10.1007/978-3-319-28706-5_4

regular maintenance with the objective of containing public health risks in the wake of flooding, especially in urban slums.

Even in a globalized economy, national governments and local administrations play a crucial role in disaster risk management. This is especially the case where droughts and floods can adversely affect livelihoods. Based on Hyogo Framework for Action (HFA) Monitor data,[1] which reports on self-assessments done by national, regional and local governments against the five priorities of the HFA, the United Nations Office for Disaster Risk Reduction (UNISDR) (2013: 212) acknowledges that governments significantly improved disaster response and preparedness strategies, but that the progress in addressing the underlying risks, through which disaster risk accumulates, is much smaller. Since prospective risk management remains a challenge for most governments, it is important to direct efforts to anticipate risks both in public and in private investments. The report calls for a new paradigm for disaster risk governance that includes both the public and the private sector. The creation of incentives for risk sensitive investment may contribute to this strategy.

Disaster risk governance is a function of institutions at multiple levels of government to predict and effectively respond to threats posed by global changes such as urbanization, climate and demographic change. An important determinant of government's effectiveness in responding to environmental threats could be levels of administrative and political decentralization. In Chap. 3, we discussed two case studies and examined how results-based financing (RBF) models could support a strategy for the delivery of water and sanitation services. In this chapter, we outline tentative research propositions to discuss the role of decentralization in predicting and responding to disaster risk. We also outline the applications of a Nexus Observatory by discussing the role of the following tools: nexus index, data visualization, scenario analysis, and benchmarking. We argue that the above tools could support the use of RBF approaches that strengthen accountability in revenue and expenditure decisions of local authorities.

2 Flood and Drought Definitions and Typologies

Floods occur when a body of water (river, lake) overflows its normal confines due to rising water levels or when there is an abnormal accumulation of water on the surface due to excess rainfall and rise of the groundwater table above the surface.

---

[1]Since 2007, national governments have been self-assessing their progress against the five priorities outlined by the HFA: (1) governance: organizational, legal and policy frameworks; (2) risk identification, assessment, monitoring and early warning; (3) knowledge management and education; (4) reducing underlying risk factors; and (5) preparedness for effective response and recovery. In 2009, a regional self-assessment process was established and, in 2011, a similar process began for local governments. The HFA review process is voluntary and intends to stimulate an inter-disciplinary planning process of disaster risk. The HFA Monitor is an online tool, facilitated by UNISDR, which allows for comparisons of data over time and across countries on progress regarding indicators of HFA.

Inundation by melting snow and ice, backwater effects, and special causes such as the outburst of a glacial lake or the breaching of a dam are also considered as floods according to the CRED glossary.[2]

Flooding is one of the most frequent natural hazards in the world. In recent decades, the damage caused by floods has been extremely severe, both in terms of human and economic losses (Douben 2006). According to Jha et al. (2011), the number and scale impact of flood events will probably accelerate in the next 50 years. Rapid urbanization, especially in low and middle-income developing countries, led to bad planning, which increased vulnerability to floods. Climate change associated with temperature increase, rise in sea levels, and more frequent extreme events will increase exposure to flooding. Therefore, flooding results from a combination of meteorological and hydrological factors but can also be induced by human activities. Table 1 presents a typology of floods according to their underlying causes.

Drought can be defined as a "deficiency of precipitation over an extended period of time, usually a season or more, which results in a water shortage for some activity, group, or environmental sectors" (UNISDR 2009a: 8). It is a temporary phenomenon, and should not be confused with aridity, which is a permanent feature of climate. Drought differs from other natural hazards, such as floods, earthquakes, and landslides, in several ways. In contrast to other hazards, it is difficult to determine the onset and end of the event and its duration is larger ranging from months to years. Another distinctive feature of drought is that its impacts are diffuse and spread slowly over a large geographical area, are difficult to quantify, and are cumulative over time.

Droughts can be analyzed from multiple disciplinary perspectives, leading to several definitions of drought.[3] Meteorological drought is defined by a precipitation deficiency over a pre-determined period of time. When there is insufficient soil water to support crop and forage growth an agricultural drought occurs. Hydrological drought results from shortages in surface and subsurface water supplies relative to some average condition at various points in time. Socio-economic drought results from shortage of supply relative to demand on some commodity or economic good that depends on precipitation (UNISDR 2009a). A socio-economic drought will not occur without one or more of the other droughts, except when societal demand consistently exceeds natural supply (Keyantash and Dracup 2002).

Figure 1 illustrates the relationship between the different types of drought. Agricultural, hydrological, and socio-economic drought are not as frequent as meteorological drought since they focus on the interaction between the natural characteristics of meteorological drought and human activities that depend on the availability of water supplies. The direct relationship between the various types of

---

[2]Center for Research on Epidemiology for Disaster (CRED): http://www.emdat.be/glossary/9.

[3]According to Wilhite et al. (2014), the inexistence of a universally accepted definition of drought leads to confusion about its real existence and degree of severity.

Table 1 Types and causes of floods

Types of flooding	Causes		Onset time	Duration
	Naturally occurring	Human induced		
Urban flood	Fluvial Coastal Flash Pluvial Groundwater	Saturation of drainage and sewage capacity Lack of permeability due to increased concretization Faulty drainage system and lack of management	Varies depending on the cause	From few hours to days
Pluvial and overland flood	Convective thunderstorms, severe rainfall, breakage of ice jam, glacial lake burst, earthquakes resulting in landslides	Land used changes, urbanization Increase in surface runoff	Varies	Varies depending upon prior conditions
Coastal (Tsunamis, storm surge)	Earthquakes Submarine volcanic eruptions Subsidence coastal erosion	Development of coastal zones Destruction of coastal natural flora (e.g. mangrove)	Varies but usually fairly rapid	Usually a short time however sometimes takes a long time to recede
Groundwater	High water table level combined with heavy rainfall Embedded effect	Development in low-lying areas; interference with natural aquifers	Usually slow	Longer duration
Flash flood	Can be caused by river, pluvial or coastal systems; convective thunderstorms; GLOFs	Catastrophic failure of water retaining structures Inadequate drainage infrastructure	Rapid	Usually short Often just a few hours
Semi-permanent flooding	Sea level rise, land subsidence	Drainage overload, failure of systems, inappropriate urban development Poor groundwater management	Usually slow	Long duration or permanent

Source World Bank (2012: 56)

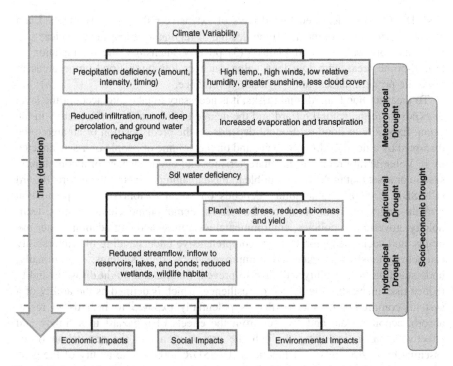

Fig. 1 Relationship between meteorological, agricultural, hydrological and socio-economic drought. *Source* UNISDR (2009a: 9)

drought and precipitation shortages decreases over time since water availability is influenced by the management of surface and subsurface water systems. For example, the use of drought-resistant crops can reduce the effects of drought.

3 Flood and Drought Risk

Flood or drought risk can be defined in a similar way as disaster risk. According to the 2009 UNISDR Terminology, disaster risk is the potential disaster losses, in lives, health status, livelihoods, assets, and services, which could occur to a particular community or a society over some specified future time period. Accordingly, risk reduction is defined as the concept and practice of reducing disaster risks through systematic efforts to analyze and manage the causal factors of disasters, including through reduced exposure to hazards, lessened vulnerability of people and property, wise management of land and the environment, and improved preparedness for adverse events (UNISDR 2009c: 10).

Extensive risk is associated with localized, mainly weather-related hazards with short return periods (UNISDR 2013: 68), such as flooding and drought.

UNISDR (2009b) identified four drivers of extensive risk: (1) badly planned and managed urban development; (2) the decline of regulatory ecosystem services (i.e. wetlands, forests, and flood plains); (3) poverty intensifies the urbanization of hazard-prone areas with low land value, such as low-lying flood-prone areas or landslide-prone hillsides; and (4) weak governance.

To measure flood and drought risks, it is necessary to take into account the level of exposure to the natural hazard and the degree of vulnerability of a society to the event. Exposure to flood or drought depends on the predisposition of an area to be disrupted by a flood or drought event and on the volume of people or assets affected by it. Vulnerability refers to the characteristics and circumstances of a community, system or asset that make it susceptible to the damaging effects of the natural hazard (UNISDR 2009c).[4] Vulnerability depends on social factors such as population growth, migration from rural to urban areas, demographic characteristics, technology, government policies, environmental awareness and degradation, water use trends, and social behavior. A more comprehensive understanding of vulnerability can help regions to anticipate and to improve early warning systems.[5] Additionally, understanding vulnerability will help to prevent flood and drought disasters and to reduce risk, increasing the degree of resilience, which is defined as the ability of a system, community or society that is potentially exposed to hazards to resist, absorb, accommodate and recover from the effects of a hazard in a timely and effective manner, including through the preservation and the restoration of its essential basic structures and functions (UNISDR 2009c). The ability of the government to provide timely relief, invest in reconstruction, and buffer economic downturns heavily influences a country's economic resilience.

Disaster governance is still an emerging concept in the disaster research literature. According to UNISDR (2013), it refers to how public and private institutions and civil society coordinate in communities and on different national levels to manage and reduce disaster and climate-related risks. Based on a literature review focused on local governance for disaster risk reduction, Rao (2013) points out some key factors for an effective disaster risk governance: decentralization and capacity building of local governments, communities, and networks to manage disaster risk; decentralization for disaster risk management as a way to guarantee efficiency, social participation, and accountability; and the political commitment from both local and national authorities. Consultations and information that local governments self-reported to the UN Office for Disaster Risk Reduction about what they consider

---

[4]According to the 2014 report of the Intergovernmental Panel on Climate Change, vulnerability is the propensity or predisposition to be adversely affected. Vulnerability encompasses a variety of concepts and elements including sensitivity or susceptibility to harm and lack of capacity to cope and adapt.

[5]Improvements in transport and health facilities, namely in East Asia and the Pacific, facilitated evacuation plans and medical assistance, reducing vulnerability, particularly in the case of flood and tropical cyclones, although exposed population increased. However, in areas where increased exposure was not accompanied by measures to reduce vulnerability, i.e. sub-Saharan Africa, flood mortality has been growing since 1980 (UNISDR 2011).

to be the most important conditions for successful risk reduction, clearly indicate that the institutional and administrative frameworks are the most important and essential elements for effective risk reduction (UNISDR 2013).

4 A Review of Existing Flood and Risk Indices

Natural hazards like floods and droughts are commonly associated with climate vulnerability. Measures of vulnerability have been first developed in the context of climate change studies. Yohe and Tol (2002) proposed a method for evaluating systems' abilities to handle external stress that lie beyond their control, namely those associated with climate change. Brenkert and Malone (2005) emphasized the importance of changing the focus from descriptive impacts of climate or climate changes to quantitative analysis of vulnerability and resilience. They constructed a vulnerability-resilience indicator prototype (VRIP), for India and Indian states, which takes into account indicators of demographic characteristics, economics, politics/governance, management of natural resources, and civil society. The VRIP proposed by Brenkert and Malone (2005) was used by Yohe et al. (2006) as a proxy to adaptive capacity. The adaptive capacity index, combined with a measure of climate change exposure, was used to assess the distributions of vulnerability to climate change across the globe. A formal mathematical framework of vulnerability was proposed by Ionescu et al. (2009). They derived an index of adaptive capacity of society to prepare for and respond to impacts of climate change by choosing an appropriate action using data on GDP per capita, literacy rate, and labor participation rate of women.

International organizations, such as the United Nations and the World Bank, have also carried out studies to assess natural hazards risk.

1. The Bureau for Crisis Prevention and Recovery of the United Nations Development Programme (UNDP) developed a disaster risk index (DRI) to measure the relative vulnerability of countries to four natural hazards: earthquakes, tropical cyclones, floods and droughts (UNDP 2004). The risk refers essentially to potential loss of life and is conceived as a function of physical exposure and human vulnerability, resulting from physical, social, economic, and environmental factors. Econometric methods were used to study how physical exposure to a natural hazard and a set of more than 20 vulnerability indicators help explain actual losses of human lives in each country during the period 1980–2000. For floods, the only vulnerability indicators, which turned out statistically significant were Gross Domestic Product (GDP) per capita and local density of population. For drought, only the percentage of population with access to improved water supply seemed to matter. The report acknowledges several main difficulties in modelling floods and droughts. First, the use of watersheds affected by floods to delimit hazard exaggerates the extent of flood-prone areas, overemphasizing human exposure and diminishing vulnerability measures. Second, in the absence of historical data on floods, annual probabilities of flood

occurrence should be based on hydrological models rather than inferred from the flood entries in the Emergency Events Database EM-DAT. Third, deaths are a limited representation of drought risk, and there are measurement errors in the recorded number of deaths (which, in many cases, may in fact be due to armed conflict).

2. The World Bank, through the natural Disaster Hotspots project, assessed the global risk in terms of two outcomes related to disaster (mortality and GDP losses), of six natural hazards: earthquakes, volcanos, landslides, floods, droughts, and cyclones, as well as for all hazards combined (Dilley et al. 2005).

3. The Bündnis Entwicklung Hilft (Alliance Development Works), in cooperation with the Institute for Environment and Human Security of the United Nations University (UNU-EHS) has been publishing, since 2011, the *World Risk Report*. Their World Risk Index (WRI) takes into account exposure to natural hazards (earthquakes, storms, floods, droughts, and sea level rise) and the vulnerability of the population, that is, its susceptibility and capacities to cope with and to adapt to future natural events. Susceptibility (likelihood of suffering harm) and coping and adaptation capacities are functions of a series of socio-economic indicators related to public infrastructure, nutrition, poverty, income, governance, medical services, education, gender equality, etc. (Bündnis Entwicklung Hilft 2014).

4. The Munich Re Group (2004) created the Hazard Index for Mega-cities (HIM), which is geared toward the risk of material losses and takes into account hazard, vulnerability, and exposed values in megacities. The hazards considered were earthquakes, windstorms, floods, volcanic eruptions, bush fires, and winter damage (frost). The risk of flood was found to be particularly high in Calcutta (India) and in Dhaka (Bangladesh).

5. Finally, DARA (2011) (an independent non-profit organization) developed a Risk Reduction Index for Central America and for West Africa (DARA 2013). They compute national and local level indices. The former are based on data available from international datasets, focusing on four types of risk drivers (environmental and natural resources, socio-economic conditions and livelihoods, land use and built environment, and governance). The latter are based on questionnaires that try to capture perceptions of risk-related conditions and capacities at the local level.

Studies including socio-economic and institutional data for vulnerability analysis in the specific context of flood or drought risk appeared more recently and are still relatively scarce. Balica et al. (2009) describe and apply a methodology for using indicators to compute a flood vulnerability index for various special scales: river basin, sub-catchment, and urban areas. Okazawa et al. (2011) estimate potential future flood damage by type of flood (heavy rain, brief torrential rain, monsoonal rain, tropical cyclone). For flood vulnerability indices in urban areas see Nasiri and Shahmohammadi-Kalalagh (2013) and Balica et al. (2012) for nine coastal cities. Finally, the International Centre for Water Hazard and Risk Management (ICHARM) and Public Works Research Institute (PWRI) Japan developed flood disaster pre-paredness indices and United Nations Educational, Scientific and Cultural Organization—Institute for Water and Education (UNESCO-IHE)[6] reports flood vulnerability indices for river basin, sub-catchment, urban areas, and coastal cities.

Regarding drought, most existing indices are based on meteorological or hydrological data for rainfall, snowpack, streamflow, soil moisture, waterway inflow and outflow, evaporation, evapotranspiration or other water supply indica-tors, and are useful to characterize a drought and its attributes and impacts, and for water supply planners and policy makers to make decisions.[7] Despite the usefulness of these indices, which only use biophysical indicators of drought, it is important to combine them with social, economic, and institutional indicators in order to capture better communities' vulnerability to drought.[8] Building on Iglesias et al. (2007), Naumann et al. (2014) proposed a drought vulnerability indicator that considers four components (renewable natural capital, economic capacity, human and civic resources, and infrastructure and technology) and 21 proxy variables to capture them. The indicator was applied at the Pan-African scale. Eriyagama et al. (2009) presented a global map of drought using several alternative indices, which take into account meteorological, hydrological, and social drought risks, as well as indices related to water infrastructure. The socio-economic drought vulnerability index they suggest is based on three indicators: the percentage contribution from agriculture to national GDP, the percentage employed in agriculture, and a crops diversity index.

None of the flood and drought vulnerability indices surveyed above takes into account institutional and decentralization aspects. However, governments play a crucial role in effective disaster risk reduction. Furthermore, several studies rely on indicators of vulnerability collected from international databases, and some of them are only available at the country level. However, regions and communities of a country are frequently quite different. To improve the accuracy of evaluations, it is important to collect information at a regional or even local community level. This would address the issue of scale, an issue we examined in earlier chapters of this volume.

5 Comprehensive Flood and Drought Risk Indices

Reducing social vulnerability to flood and drought begins with policy makers and public awareness of the need to prevent and adapt to these problems. A movement from a crisis/disaster management approach to a more proactive and risk

---

[6]For more information, please refer to their website at http://unescoihefvi.free.fr/vulnerability.php.

[7]Zargar et al. (2011) reviews 74 operational meteorological, agricultural, and hydrological drought indices used for forecasting, monitoring and planning for drought.

[8]As we have argued in Chap. 3, data from various sources, including from earth observations could make synergies and trade-offs more explicit, thereby, drawing a more holistic and comprehensive picture.

management approach should clearly be part of the political reform agenda. Since risk management starts with an assessment of the risks involved, it is important to develop comprehensive risk indices that properly reflect countries' or regions' exposure and vulnerability to floods and droughts. The theoretical framework and previous studies described above provide the basis for the selection and combination of the relevant variables, and it is necessary to clearly understand and define the determinants of flood and drought risk.

As suggested in UNISDR (2009b), estimates of the risks associated with a natural hazard in an area of interest can be obtained by combining measures of exposure of people or assets located in that area with the specific vulnerability of the exposed elements to the hazard being considered. Accordingly, a comprehensive flood or drought risk index should combine measures of physical exposure to floods or droughts with indicators of vulnerability, that is, with the characteristics and circumstances of a community or asset that make it susceptible to the damaging effects of the hazard. A community's resilience or its coping and adaptation capacities will also affect the way in which exposure translates into losses of lives or assets. Thus, indicators or indices of vulnerability should also include variables that reflect coping and adaptation capacities. Given this phenomenon's multidimensionality, natural, environmental, social, economic, and institutional features of vulnerability should be taken into account.

The methodology proposed in this study to assess flood/drought risk involves identifying variables that reflect the exposure and vulnerability of people and/or assets to the natural hazard, collecting data for the countries under analysis, validating the model so that indicators are properly weighted, implementing the index in a way that increases public awareness and preparedness, and continuously evaluating the index in order to improve its predictive abilities and usefulness for policy-making.

Following UNDP (2004) and UNISDR (2009b), flood/drought risk is modelled as a function of the probability that the natural hazard occurs, the element at risk (population or assets) and vulnerability. Thus, when considering population, flood/drought risk in country or region i would be given by:

$$Risk_i = Prob(Natural\ hazard)_i \times Population_i \times Vulnerability_i \qquad (1)$$

This definition implies that flood/drought risk is zero in a geographical area where no floods/droughts occur, nobody lives or where the population is invulnerable to floods/droughts. The probability of flood/drought occurrence in a specific geographical area can be calculated by dividing the number of floods/droughts that occurred there in the past by the available years of observations.

5.1 Exposure to Floods or Droughts

A measure of the number of people exposed to the natural hazard in a given geographical area can be obtained multiplying the probability of the natural hazard occurrence in that area by its population. This measure of physical exposure is given by:

$$Physical\ Exposure_i = Prob(Natural\ hazard)_i \times Population_i \qquad (2)$$

It may also be useful to obtain a measure of economic exposure, in order to have an idea of the economic value at risk. In that case, the geographical area's population can be replaced by its Gross Domestic Product (GDP). We then obtain:

$$Economic\ Exposure_i = Prob(Natural\ hazard)_i \times GDP_i \qquad (3)$$

Although data on population and GDP can easily be obtained, measuring the probability of flood or drought occurrence is not a trivial issue. First, as illustrated in Table 1 and Fig. 1, there are several definitions of flood and drought. Second, there are several indices available to measure the probability of flood or drought occurrence. Important factors to take into account when choosing the flood and drought definition and index to use are the availability of data and the complexity of the necessary calculations. Readily and publicly available global data on physical and economic exposure to floods and droughts is provided by the *PREVIEW Global Risk Data Platform*[9] of UNEP/Global Resource Information Database (GRID-Geneva) and UNISDR.[10] One disadvantage of this exposure data is that the drought index used, the Standardized Precipitation Index (SPI), is only based on precipitation data. Although the SPI was considered and recommended as the universal best meteorological drought index by the Inter-Regional Workshop on Indices and Early Warning Systems for Drought (Hayes et al. 2011), meteorological drought does not necessarily imply the occurrence of the other types of drought. Thus, this dataset provides an imperfect measure of actual exposure to drought.

---

[9]For more information on the PREVIEW Global Risk Data Platform, please refer to http://preview.grid.unep.ch/.

[10]The physical exposure measure is the expected average annual population (2007 as the year of reference) exposed (inhabitants) to floods/droughts. The estimate of the annual physical exposition to flood is based on four sources: (1) A GIS modelling using a statistical estimation of peak-flow magnitude and a hydrological model using HydroSHEDS dataset and the Manning equation to estimate river stage for the calculated discharge value; (2) Observed flood from 1999 to 2007, obtained from the Dartmouth Flood Observatory (DFO); (3) The frequency was set using the frequency from UNEP/GRID-Europe PREVIEW flood dataset. In areas where no information was available, it was set to 50 years returning period; (4) A population grid for the year 2010, provided by LandScanTM Global Population Database (Oak Ridge, TN: Oak Ridge National Laboratory). For drought, the dataset includes an estimate of global drought annual repartition based on the Standardized Precipitation Index, and is based on three sources: (1) A global monthly gridded precipitation dataset obtained from the Climatic Research Unit (University of East Anglia); (2) A GIS modelling of global Standardized Precipitation Index based on Brad Lyon (IRI, Columbia University) methodology; (3) A population grid for the year 2007, provided by LandScanTM Global Population Database (Oak Ridge, TN: Oak Ridge National Laboratory). In the case of economic exposure, population is replaced by GDP (US$, year 2000 equivalent), obtained from the World Bank.

Nevertheless, considering the geographical (grid) level of detail and country coverage, it is eventually the best global dataset on drought and flood exposure currently available.[11]

5.2 Vulnerability to Floods and Droughts

As stated above, a community's vulnerability to floods and droughts is a function of the likelihood that it suffers harm from flood and drought episodes (its susceptibility) and of its capacity to cope with and adapt to them. Previous studies and reports have identified a wide group of biophysical, socio-economic, and institutional characteristics of countries and/or regions that affect their vulnerability to floods and droughts (see, among others, UNDP 2004; Adler and Vincent 2005; Brenkert and Malone 2005; Balica et al. 2009, 2012; Okazawa et al. 2011; Bündnis Entwicklung Hilft 2014; and Singh et al. 2014).

Susceptibility results from community's characteristics or circumstances that affect the damage caused by floods and droughts. These encompass biophysical characteristics, economic capacity, health and sanitation services, and infrastructure and technology. Given the fragile situation of refugees and of other displaced populations, they also affect a territory's susceptibility to floods or droughts. A list of vulnerability indicators related to the abovementioned dimensions of susceptibility, which have been used in previous studies (Okazawa et al. 2011; Balica et al. 2012; DARA 2013; Bündnis Entwicklung Hilft 2014) and are available for many countries, is presented in Table 2 for floods and in Table 3 for droughts. Although most of the sources indicated only contain data at the national level, it may be possible to obtain regional/local data from national statistical agencies.

A community's vulnerability is also a function of its coping and adaptation capacities. These can be represented by several socio-economic and institutional indicators (see Table 4), which should be considered in a flood/drought risk index (UNDP 2004; Okazawa et al. 2011; Balica et al. 2012; DARA 2013; Naumann et al. 2014; Bündnis Entwicklung Hilft 2014). Regarding socio-economic indicators, greater human development, expressed in higher life expectancy and more educated communities, provides greater capacity to cope with and adapt to the natural hazard. Access to better medical services and more efficient early warning systems will also help people resist floods or droughts and decrease the number of fatalities caused by them.

---

[11]National disaster loss data for 72 countries is available from the Desinventar (http://www. desinventar.net) Disaster Information Management System. In case studies focusing on just one, or on a very small number of countries, it may be possible to collect more detailed data and to use more comprehensive drought indices than the SPI, which would then allow the construction of more precise measures of exposure to floods and droughts.

Table 2 Vulnerability indicators related to susceptibility to floods

Component	Dimension	Indicator	Data Sources	Data coverage
Biophysical	Land use and population location	Population within 100 km of coast (% of total)	UNEP/DEWA/GRID	
		Population living in areas of elevation below 5 m (% of total)	World Bank, WDI	1960–2013
		Agricultural area (% of total)	World Bank, WDI	1960–2013
		Forest area (% of total)	World Bank, WDI	1960–2013
		Population density (inhabit km^2)	World Bank, WDI	1960–2013
	Pressure on resources	Annual freshwater withdrawals, total (% of internal resources)	World Bank, WDI	1960–2013
		Human-Induced Soil Degradation (GLASOD)	FAO/UNEP	
		Urban population growth	World Bank, WDI	1960–2013
		Rural population growth	World Bank, WDI	1960–2013
Socio-economic	Economic capacity	GDP per capita USD	World Bank, WDI	1960–2013
		Population living below US$ 1.25 PPP per day (%)	World Bank, WDI	1960–2013
		Age dependency ratio	World Bank, WDI	1960–2013
	Health and sanitation	Notified cases of malaria (per 100,000 people)	World Bank, WDI	1960–2013
	Human displacement	Displaced persons	Internal Displacement Monitoring Center	2008–2013
	Infrastructure and technology	% of people living in slums	MDG Website	
		Roads paved (% of total roads)	World Bank, WDI	1960–2013
		Number of dams	University of Yamanashi	
		Maximum reservoir storage (1000 m^3)	University of Yamanashi	

Note: *WDI* World Development Indicators

Table 3 Vulnerability indicators related to susceptibility to drought

Component	Dimension	Indicator	Data sources	Data coverage
Biophysical	Water management	Agricultural water use (% of total)	Aquastat	98-02, 03-07, 08-12, 13-17
		Annual freshwater withdrawals, total (% of internal resources)	World Bank, WDI	1960–2013
		Agricultural irrigated land (% of total agricultural land)	World Bank, WDI	1960–2013
•		Agricultural land (% of total area)	World Bank, WDI	1960–2013
	Pressure on resources	Population density (inhabit km^{-2})	Aquastat	98-02, 03-07, 08-12, 13-17
		Human-Induced Soil Degradation (GLASOD)	FAO/UNEP	
Socio-economic	Economic capacity	GDP per capita USD	World Bank, WDI	1960–2013
		Population living below USD 1.25 PPP per day (% of total population)	World Bank, WDI	1960–2013
		Age dependency ratio	World Bank, WDI	1960–2013
		Fertilizer consumption (kg per ha of arable land)	World Bank, WDI	1960–2013
	Economic structure	Agricultural value added (% of GDP)	World Bank, WDI	1960–2013
		Labor force in agricultural sector (% of total)	World Bank, WDI	1960–2013
	Health and sanitation	% of people with access to improved water supply	WHO/UNICEF; WB, WDI	1960–2013
		Proportion of undernourished people	World Bank, WDI	1960–2013
	Human displacement	Displaced persons (% of total population)	Internal Displacement Monitoring Center	2008–2013
	Infrastructure and technology	Water infrastructure (storage as proportion of total renewable water resources)	Aquastat	98-02, 03-07, 08-12, 13-17 (5-year data)
		Energy use (kg oil equivalent per capita)	World Bank, WDI	1960–2013

Note: *WDI* World Development Indicators

Table 4 Vulnerability indicators related to coping and adaptive capacities

Component	Dimension	Indicator	Data sources	Data coverage
Socio-economic	Human development	Adult literacy rate (%)	World Bank, WDI	1960–2013
		Life expectancy at birth (years)	World Bank, WDI	1960–2013
		School enrollment ratios	World Bank, WDI	1960–2013
		Human development index	UNDP	1980–2013
	Medical services	Number of physicians (per 1000 people)	World Bank, WDI	1960–2013
		Number of hospital beds (per 1000 inhabitants)	World Bank, WDI	1960–2013
	Early warning systems (flood index only)	Number of mobile cellular subscriptions (per 100 people)	World Bank, WDI	1984–2013
		Internet users (per 100 people)	World Bank, WDI	1990–2013
		Number of radios (per 1000 people)	CNTS	1900–2012
Institutional	Governance	Control of corruption	World Bank, WGI	1996–2013
		Government effectiveness	World Bank, WGI	1996–2013
		Rule of law	World Bank, WGI	1996–2013
		Law and order	World Bank, WGI	1996–2013
		Percentage of women in Parliament	World Bank, WDI	1960–2013
	Political stability	Political stability	World Bank, WGI	1996–2013
		Civil wars	SFTF	1815–2012
		Government crises	CNTS	1815–2012
		Revolutions	CNTS	1815–2012
		Coups d'état	CNTS	1815–2012
	Decentralization	Subnational gov. expenditure (% total gov. expenditure)	Government Financial Statistics, IMF	1995–2012
		Transfers from central government (% of revenues of sub-national governments)	Government Financial Statistics, IMF	1995–2012
		Locally elected sub-national governments	DPI, World Bank	1975–2012

Notes: *WDI* World Development Indicators; *WGI* Worldwide Governance Indicators; *CNTS* Cross National Time Series database; *SFTF* State Failure Task Force dataset; *DPI* Database of Political Institutions

Several institutional factors affect a community's coping and adaptation capacities, namely the quality of governance, political stability, and degree of decentralization. UNDP (2004) argues that good governance is key for sustainable development and disaster risk reduction. In fact, weak governance is one of the most important risk factors regarding the impact of natural hazards, as states with strong institutions have fewer deaths after extreme natural events than those with weak governance (Bündnis Entwicklung Hilft 2011). In the latter, governments are rarely able or ready to organize and implement an efficient system of disaster preparedness. The quality of governance can be assessed using the Worldwide Governance Indicators (Kaufmann et al. 2010). Women participation in politics is another institutional aspect to consider.

Political stability is also a very important determinant of a community's vulnerability to flood and drought. In the midst of civil wars or of generalized popular unrest, it is practically impossible for the government to protect its citizens effectively from natural hazards or to assist them after they occur. According to UNDP (2004), it is possible that part of the deaths recorded in the EM-DAT database, especially in African countries, are in fact due to armed conflict. Even mild manifestations of political instability, such as high turnover of governments, can interfere with effective flood and drought risk management and result in higher human and material losses. Thus, indicators of political instability should be considered when constructing a natural hazard risk index.

Given that the frequency and severity of flood and drought may vary substantially from one country's region to another, decentralization, by attributing greater competencies to sub-national governments, may also help improve risk management. Through a bottom-up approach, local concerns related to floods, droughts, and factors that influence vulnerability/resilience might become more visible to the national authorities. Community participation and good coordination of organizations and institutions at all levels are key to the design and implementation of effective policies and strategies to reduce disaster risk. Local governments are closer to citizens and, therefore, are better able to understand the needs of the population and are capable of designing policies that satisfy their needs, develop coping capabilities, and increase their safety. They also play an important role in informing central governments of local population needs. The recognition of decentralization as an important element of good governance is also present in the International Guidelines on Decentralization and Access to Basic Services for all (UN-HABITAT 2009).

The inclusion of decentralization indicators in a flood or drought risk index is relevant, as high degrees of discretion (decision-making autonomy) will allow local governments to react and even anticipate disaster risk in a specific area. It is, therefore reasonable to admit that decentralization has a positive impact on: the resilience of food systems, agricultural yields, and nutritional security (disaggregated by gender, class, and ethnicity). The existence of locally elected sub-national governments and the share of sub-national governments' expenditure in total public expenditure are used as indicators of decentralization. The degree of sub-national governments' dependence from intergovernmental transfers may negatively influence their performance.[12]

5.3 Normalization, Weighting and Validation

As pointed out by the Organisation for Economic Co-operation and Development (OECD) (2008), we also need to be aware of the normalization, weighting, and aggregation procedures. The normalization of variables to a common baseline is necessary, especially when comparing results, as indicators have different measurement units. One method presented in OECD (2008) is the Min-Max normalization. This method normalizes indicators to have an identical range between 0 and 1. For positive correlations to risk, the normalized value is calculated as:

$$Z_i = \frac{X_i - X_{\min}}{X_{\max} - X_{\min}} \tag{4}$$

where X_i represents the variable's value for a country, i, $X_{\min}$ and $X_{\max}$ are the respective minimum and maximum values across all countries. For negative correlations, we have:

$$Z_i = 1 - \left(\frac{X_i - X_{\min}}{X_{\max} - X_{\min}}\right). \tag{5}$$

As a result, it is possible to compute a mean of the normalized variables that define each component:

$$C_K = \frac{1}{n}\sum_{K=1}^{n} Z_k \tag{6}$$

The selection of the weighting scheme may have a significant effect on the value of the index and this may alter country rankings. The use of equal weighting is extensive in the construction of composite indicators but it can be the case that "there is insufficient knowledge of causal relationships or lack of consensus on the alternative" (OECD 2008: 31). In the case of variables that are grouped into

---

[12]A wide portfolio of revenue sources improves governments' risk sharing and helps to handle the impact of unexpected events in revenues. Furthermore, when sub-national governments are highly dependent on transfers, they tend to be more fiscally irresponsible. For a discussion on decentralization, see Veiga et al. (2015).

dimensions and then aggregated into the composite, the application of equal weighting will create problems in the construction of the index, as some indicators or dimensions may have greater importance than others. This concern on selecting the appropriate weighting and aggregation method is patent in Bündnis Entwicklung Hilft (2011), which uses factors analysis to validate the aggregation formula of the *World Risk Index*, and a comprehensive sensitivity analysis to examine the sources of variation in the model's output.

Alternative/complementary methodologies to attribute weights to the indicators require experts with extensive field experience dealing with floods and droughts and the damages they cause, and to use regression analysis to identify the indicators that explain better flood/drought risk. This can be done by replacing flood/drought risk in Eq. (1) by its observed effects, such as deaths or economic losses attributed to floods/droughts, and then regress them on flood/drought exposure and vulnerability indicators.[13] The estimated model for deaths caused by floods can be represented as follows[14]:

$$Deaths_i = Physical\ Exposure_i \times Vulnerability_i \tag{7}$$

Taking natural logarithms on both sides, we obtain a log-linear model, which can be easily estimated with most of the available statistical software packages:

$$\ln(Deaths)_i = \ln(Physical\ Exposure)_i + \ln(Vulnerability)_i \tag{8}$$

This regression analysis may then be used to improve the proposed flood/drought risk index. Keeping only the statistically significant indicators and using the estimated coefficients as a reference for the weighting scheme, a more parsimonious model would be obtained and no ad hoc assumptions regarding weights (equal, proportional or random) would be used. Thus, more precise risk quantification could be carried out.[15]

---

[13]That was the procedure adopted in UNDP (2004), where risk, assessed by the number of deaths caused by a natural hazard (using data from *EM-DAT*), was then regressed on exposure and on several vulnerability indicators. A similar methodology was also applied by UNISDR (2009b, 2011, 2013), Yohe and Tol (2002) and Okazawa et al. (2011).

[14]The determinants of economic losses caused by floods or droughts could be investigated by replacing "deaths" with "economic losses" (from *EM-DAT* or *Desinventar*) and "physical exposure" with "economic exposure." It is worth noting that "vulnerability" does not represent one single variable, but a vector of indicators of susceptibility (Tables 2 and 3) and coping and adaptation capacities (Table 4).

[15]An estimate of the global risk induced by flood hazard is available from the *PREVIEW Global Risk Data Platform*. This index, used in GAR13 (UNISDR 2013), varies from 1 (not null) to 10 (extreme). A new, probabilistic risk assessment approach is currently under way and the new flood risk index should be made available in the GAR15. Its key output is the likelihood of having certain losses from floods expressed in terms of their occurrence rate, expressed per year.

6 The Nexus Observatory, Governance and Disaster Risk Management[16]

The response, preparedness, and development of improved flood and drought monitoring, early warning systems and information diffusion to policy makers with respect to flood and drought is still too little and too late. The capacity to measure flood and drought risks accurately and to predict future hazards is essential to reduce communities' vulnerability to floods, droughts, and food insecurity. For that purpose, the adoption and constant improvement of comprehensive flood and drought risk indices is essential. Although it is usually much harder to collect data at the sub-national level, flood and drought risk assessment should also be conducted at the local level, as flood exposure and vulnerability can vary considerably from one country's region to another.[17]

These measures may help policy makers in national government institutions and international organizations to assign funds to projects that reduce the vulnerability of communities, green taxes can be designed to create incentives for firms to be environmentally friendly, and intergovernmental transfers may also be targeted to regions with a larger risk of a natural hazard to address the underlying drivers of risk. Additionally, land-use planning and management should be used to encourage effective disaster risk management, and a more pro-active role of the private sector should be encouraged. Local governments can find allies among businesses with large fixed assets in the municipality to implement a more efficient risk management. According to Kataria and Zerjav (2012), "communities of interest" are being created by local governments and businesses, which can contribute to the reduction of vulnerability. Disaster risk metrics should be integrated into private and public investment and planning and, easy access to risk information and up-to-date estimates are necessary.

The UNU-FLORES Nexus Observatory intends to foster informed decisions toward sustainable development with regards to the nexus of water, soil, and waste. Towards this end, the Nexus Observatory focusses on an important relationship that could potentially connect the science and policy-making domains. This relationship relates to poverty and its effect on the environment. It is commonly believed that poverty (*lower incomes or limited entitlements to natural resources*) leads people to over-extract environmental resources (Bebbington 1999). Over-extraction of environmental resources could take the form of excessive harvesting of forest resources, over-pumping of groundwater or intensive cropping regimes. Conversely, one may argue that improved management of environmental resources may result in reduction in incidence of poverty. The causality and direction of the relationship between poverty and condition of environmental resources is a contested one as

---

[16]See Chap. 3 for a detailed account of the benefit of using web-based data observatories.

[17]For applications of indices to regions of specific countries see, among others, Bündnis Entwicklung Hilft (2011) for Indonesia, UNISDR (2013) for several individual countries, DARA (2013) for six West African Countries, and DARA (2011) for seven Central American countries.

borne out by research on biodiversity, air quality or water quality (see Dasgupta et al. 2005). The relationship between poverty and the condition of environmental resources is critical in determining how governments (*as important stewards of crucial environmental resources*) should respond in terms of allocating scarce human and financial resources. We pointed out earlier in this Brief that where the relationship between poverty and the environment is not strong sector wide budgetary approaches may suffice. On the other hand, where the relationship between poverty and condition of environmental resources is strong budget support strategies may be more effective.

Scientific research that is relevant to policy debates, can bridge the science-policy divide by providing decision makers with tools and instruments that would enable them to make informed choices between a range of technical options, human resources, and financial models. For this reason a mechanism that clarifies the role of confounding variables, such as scale and boundary conditions is crucial. Further, important clarifications and qualifications of policy prescriptions become credible when important issues relating to data—its reliability, frequency, and dis-aggregation are addressed. We are optimistic that the Nexus Observatory by being able to effectively perform this role can go a long way in assisting decision makers in responding to environmental risks and mitigate their potential impacts on incidence of poverty. To better appreciate the importance of revenue and expenditure decisions of governments the literature on decentralization is revealing in many respects as can be seen from the ensuing discussion.

6.1   The Trend in Decentralization

In the previous chapter decentralization was examined in relation to public services. We will now turn to discussing the role of decentralization as a strategy for mitigating the effects of floods and droughts. In terms of governance, decentralization has the potential to strengthen flood and drought risk management, as regional and local authorities are closer to the people affected by floods and droughts than national governments are and should, thus, be better able to assess their exposure and vulnerability and to assist them when natural hazards occur. Greater decentralization also tends to increase community participation, which is an essential component of flood and drought risk management. It is crucial to involve the local population in flood and drought mitigation strategies, in the planning of disaster management, allocation of resources, or implementation of the plan. Decentralization can, therefore, increase the resilience of food systems, as assumed in Sect. 5.2.

Additionally, decentralization can improve agricultural yields, since greater autonomy of sub-national governments increases the ability of policy makers and extension agencies to respond to farmer's need for better seeds, fertilizers or information on crop prices. Sub-national governments may serve as a link between farming communities and higher levels of decision-making, enhancing information exchange and capacity building.[18] Finally, greater decentralization may increase nutritional security, especially of the most vulnerable individuals. As the food security of households improves, it is likely to become more evenly distributed across their members, irrespective of gender, age, class or ethnicity.

Despite the abovementioned potential, experience in many developing countries has shown that decentralization brings several challenges to disaster risk reduction, such as coordination problems, financing difficulties, and capacity constraints (Scott and Tarazona 2011). Furthermore, decentralization has not automatically increased participation, and lack of accountability of local policy makers has resulted in the diversion of disaster risk reduction funds to other uses. Given these difficulties, it is very important to improve technical capacity at the local level, to provide sub-national governments with the adequate level of financial resources, to design contracts and incentive systems that enhance sub-national governments' performance, and to promote public awareness about disaster risk.

6.2 Implications for Planning and Management of Environmental Resources

The Nexus Observatory, as was highlighted in Chap. 3 of this Brief, will serve as an interface between science and policy by disseminating relevant information and knowledge for environmental resource planning, management and by creating monitoring frameworks and tools.[19] It recognizes that decision makers should not simply be guided by findings of controlled experiments/field trials with no valid comparison group, but rather by the results of studies with strong research designs that clearly identify causal effects.

The Nexus Observatory will allow for data visualizations that are likely to: (1) stimulate stakeholders to pose questions that require policy-relevant research; (2) allow for performance monitoring through multiple indicators; (3) enhance transparent discussion of trade-offs between sometimes competing objectives of decision-making, namely efficiency and equity; and (4) facilitate collective action requiring political negotiation (e.g. among authorities at different levels) to achieve a commonly agreed goal. The construction of indices, based on and supported by data visualization tools, may facilitate political negotiations due to their transparency and quantitative nature. They have the potential to inform decisions on the allocation of financial or human resources by donors and by different layers of government for integrated management of environmental resources. Regarding flood and drought risk, visualization tools and indices are important, not only to help decision makers understand differences in disaster incidence, but also to show

---

[18]See Kurian (2010) for an analysis of soil conservation interventions in Laos and the importance of socio-ecological data for planning and management interventions.

[19]For a detailed description of the Nexus Observatory, see Kurian and Meyer (2014).

the impact of decentralization on the resilience of food systems, agricultural yields, and nutritional security.

Scenario analysis is also influential for decision makers to understand consequences of acting or not acting. Key steps involve: (1) an assessment of the current situation, (2) the definition of alternative scenarios containing broad goals under different levels of risk and assumptions, (3) the identification of the role of stakeholders and of the required monetary and human resources, and (4) specifying transparent and measurable indicators. Scenario analysis helps to build consensus on sustainable financing of collective action to achieve identifiable objectives. Benchmarking is necessary to examine over time, which governments are responding, and to enable the design of incentive schemes such as budget support. Financing and accountability problems at the different levels of government could be mitigated through the use of RBF. As argued in the previous chapter and in greater detail by Veiga et al. (2015), OBA and other RBF models bring in the funding required and strengthen accountability by service providers, including local governments. By tying the disbursement of funds to the attainment of agreed upon results/outcomes, they induce greater effort and guarantee that funds are not diverted to other purposes. Thus, RBF models should be considered in a strategy of flood and drought risk reduction.

Finally, where in situ data based on field studies is not available, or is insufficient to guide decision-making, novel data collection approaches, based on new technologies, will be explored to overcome information gaps. Remote sensing and geographic information system (GIS) have the potential to make available real-time, high-quality, and reliable data that enhances evidence-based decision-making for management of environmental resources. This data will allow for a more efficient management of crises, disasters, and emergency situations.

7 Conclusion

Managing flood and drought risk requires understanding of the natural hazards and of the human, social, economic, and environmental vulnerability of the community. Although little can be done to alter flood or drought occurrence, early warning and effective prevention systems help mitigate its negative consequences. Given that recent projections indicate that the frequency and magnitude of extreme events is increasing, and taking into account the ineffectiveness of past attempts to manage natural hazards, it is crucial to increase emphasis on risk management and to adopt policies to reduce the underlying causes of societal vulnerability, building resilience for future episodes.

The adoption of comprehensive flood and drought risk indices, which take into account the multiple aspects of natural disasters and properly account for exposure and vulnerability of people and assets to it, is an essential component of risk management strategy. Additionally, it is important to make natural hazard risk management a component of poverty reduction and sustainable development pathways, rather than an issue to be addressed at a specific point in time by specific affected regions. This requires multi-stakeholder cooperation,[20] since the commitment of national, regional and local governments and effective engagement of the communities and businesses at risk are crucial for success. The definition of a clear system of responsibilities for managing risk programs by different levels of government, and the design of fiscal mechanisms that guarantee appropriate levels of funding and incentives for good performance is also critical.

References

Adler, W. N., & Vincent, K. (2005). External geophysics, climate and environment uncertainty in adaptive capacity. *Comptes Rendus Geoscience, 337*, 399–410.

Balica, S. F., Douben, N., & Wright, N. G. (2009). Flood vulnerability indices at varying spatial scales. *Water Science and Technology, 60*(10), 2571–2580.

Balica, S. F., Wright, N. G., & van der Meulen, F. (2012). A flood vulnerability index for coastal cities and its use in assessing climate change impacts. *Natural Hazards, 64*, 73–105.

Bebbington, A. (1999). Capitals and capabilities: A framework for analysing peasant viability, rural livelihoods and poverty. *World Development, 27*, 2021–2044.

Brenkert, A. L., & Malone, E. L. (2005). Modelling vulnerability and resilience to climate change: A case study of India and Indian states. *Climatic Change, 72*, 57–102.

Bündnis Entwicklung Hilft. (2014). *World risk report 2014.* Berlin: Bündnis Entwicklung Hilft.

DARA. (2011). *Risk reduction index in Central America and the Caribbean—analysis of the capacities and conditions for disaster risk reduction.* Madrid: Fundación DARA Internacional.

DARA. (2013). *Risk reduction index in West Africa—analysis of the capacities and conditions for disaster risk reduction.* Madrid: Fundación DARA Internacional.

Dasgupta, S., Deichmann, U., Meisner, C., & Wheeler, D. (2005). Where is the poverty-environment nexus? Evidence from Cambodia, Laos PDR and Vietnam. *World Development, 33*(4), 617–638.

Dilley, M., Chen, R. S., Deichmann, U., Lerner-Lam, A., & Arnold, M. (2005). *Natural disaster hotspots: A global risk analysis.* Washington, DC: World Bank, Hazard Management Unit.

Douben, K. (2006). Characteristics of river floods and flooding: A global overview. *Irrigation and Drain, 55*, S9–S21.

Eriyagama, N., Smakhtin, V., & Gamage, N. (2009). Mapping drought patterns and impacts: A global perspective. *IWMI Research Report 133.* Colombo, Sri Lanka: International Water Management Institute.

Hayes, M., Svoboda, M., Wall, N., & Widhalm, M. (2011). *The Lincoln declaration on drought indices: Universal meteorological drought index recommended.* Boston, MA: American Meteorological Society.

Iglesias, A., Moneo, M., & Quiroga, S. (2007). Methods for evaluating social vulnerability to drought. *Options Méditerranéenes, Séries B, Etudes et Recherches, 58*, 129–133.

Ionescu, C., Klein, R. J. T., Hinkel, J., Kavi Kumar, K. S., & Klein, R. (2009). Towards a formal framework of vulnerability to climate change. *Environmental Model Assessment, 14*, 1–16.

Jha, A., Lamond, J., Bloch, R., Bhattacharya, N., Lopez, A., Papachristodoulou, N., Bird, A., Proverbs, D., Davies, J., & Barker, R. (2011). Five feet high and rising—cities and flooding in the 21st century. *Policy Research Working Paper 5648.* Washington DC: World Bank—East Asia and Pacific Region, Transport, Energy & Urban Sustainable Development Unit.

---

[20]Proposed SDG Goal 17 also highlights multi-stakeholder partnerships.

Kataria, S., & Zerjav, B. (2012). Private sector investment decisions in building and construction: Increasing managing and transferring risks. A literature review. Prepared for the *2013 Global Assessment Report on Disaster Risk Reduction*. Geneva, Switzerland: UNISDR.

Kaufmann, D., Kraay, A., & Mastruzzi, M. (2010). The worldwide governance Indicators: Methodology and analytical issues. *World Bank Policy Research Working Paper 5430*.

Keyantash, J., & Dracup, J. A. (2002). The quantification of drought: An evaluation of drought indices. *Bulletin of the American Meteorological Society, 83*(8), 1167–1180.

Kurian, M. (2010). Making sense of human-environment interaction: Policy guidance under conditions of imperfect data. In M. Kurian & P. McCarney (Eds.), *Peri-urban water and sanitation services: Policy, planning and method*. Dordrecht: Springer.

Kurian, M., & Meyer, K. (2014). *UNU-FLORES Nexus observatory flyer*. Dresden: UNU-FLORES.

Munich Re Group. (2004). *Megacities—Megarisks trends and challenges for insurance and risk management*. Munich: Munich Re Group.

Nasiri, H., & Shahmohammadi-Kalalagh, S. (2013). Flood vulnerability index as a knowledge base for flood risk assessment in urban area. *Journal of Novel Applied Science, 2*(8), 269–272.

National Drought Mitigation Center. (2014). *Drought basics*. Asheville, NC: National Climatic Data Center.

Naumann, G., Barbosa, P., Garrote, L., Iglesias, A., & Vogt, J. (2014). Exploring drought vulnerability in Africa: An indicator based analysis to be used in early warning systems. *Hydrology and Earth System Sciences, 18*, 1591–1604.

OECD. (2008). *Handbook on constructing composite indicators—methodology and user guide*. Paris: OECD.

Okazawa, Y., Yeh, P. J., Kanae, S., & Oki, T. (2011). Development of a global flood risk index based on natural and socio-economic factors. *Hydrological Sciences Journal, 56*(5), 789–804.

Rao, S. (2013). Disaster risk governance at national and sub-national levels. *GSDRC Helpdesk Research Report 991*. Birmingham, UK.

Scott, Z., & Tarazona, M. (2011). Decentralisation and disaster risk reduction. *Background report for the global assessment report on disaster risk reduction 2011*. Geneva, Switzerland: United Nations International Strategy for Disaster Reduction (UNISDR).

Singh, N. P., Bantilan, C., & Byjesh, K. (2014). Vulnerability and policy relevance to drought in the semi-arid tropics of Asia—a retrospective analysis. *Weather and Climate Extremes, 3*, 54–61.

UNDP. (2004). *Reducing disaster risk: A challenge for development*. New York: UNDP, Bureau for Crisis Prevention and Recovery.

UNDP. (2014). *Governance for sustainable development. Integrating governance in the post-2015 development framework*. Discussion paper. New York: UNDP.

UN-HABITAT. (2009). *International guidelines on decentralization and access to basic services for all*. Nairobi, Kenya: United Nations Human Settlements Programme.

UNISDR. (2009a). *Drought risk reduction framework and practices: contributing to the implementation of the Hyogo framework for action*. Geneva, Switzerland: United Nations International Strategy for Disaster Reduction (UNISDR).

UNISDR. (2009b). *Global assessment report on disaster risk reduction*. Geneva, Switzerland: United Nations International Strategy for Disaster Reduction (UNISDR).

UNISDR. (2009c). *The 2009 UNISDR terminology*. Geneva, Switzerland: United Nations International Strategy for Disaster Reduction (UNISDR).

UNISDR. (2011). *Global assessment report on disaster risk reduction*. Geneva, Switzerland: United Nations International Strategy for Disaster Reduction (UNISDR).

UNISDR. (2013). *Global assessment report on disaster risk reduction*. Geneva, Switzerland: United Nations International Strategy for Disaster Reduction (UNISDR).

Veiga, L. G., Kurian, M., & Ardakanian, R. (2015). *Intergovernmental fiscal relations—questions of accountability and autonomy.*, Springer Briefs in Environmental Science Dordrecht: UNU-Springer.

Wilhite, D. A., Sivakumar, M. V. K., & Pulwarty, R. (2014). Managing drought risk in a changing climate: The role of national drought policy. *Weather and Climate Extremes, 3*, 4–13.

World Bank. (2012). *Cities and flooding—a guide to integrated urban flood risk management for the 21st century.* Washington DC: The World Bank.

Yohe, G., & Tol, R. S. J. (2002). Indicators for social and economic coping capacity—moving toward a working definition of adaptive capacity. *Global Environmental Change, 12,* 25–40.

Yohe, G., Malone, E., Schlesinger, M., Brenkert, A., Meij, H., & Xing, X. (2006). Global distributions of vulnerability to climate change. *The Integrated Assessment Journal, Bridging Sciences & Policy, 6*(3), 35–44.

Zargar, A., Sadiq, R., Naser, B., & Khan, F. I. (2011). A review of drought indices. *Environmental Reviews, 19,* 333–349.

Erratum to: Resources, Services and Risks

Erratum to:
M. Kurian et al., *Resources, Services and Risks,*
SpringerBriefs in Environmental Science,
DOI 10.1007/978-3-319-28706-5

The original version of this book was inadvertently published with an incorrect logo and copyright holder details, it should be changed to:

The updated original online version for this book can be found at
10.1007/978-3-319-28706-5

M. Kurian et al., *Resources, Services and Risks,*
SpringerBriefs in Environmental Science, DOI 10.1007/978-3-319-28706-5_5

Printed in the United States
By Bookmasters